INTELLECTUAL PROPERTY IN THE 21ST CENTURY

PATENTS

INTERNATIONAL PROTECTION FOR SMALL BUSINESS AND THE PRIOR USER RIGHTS DEFENSE

INTELLECTUAL PROPERTY IN THE 21ST CENTURY

Additional books in this series can be found on Nova's website under the Series tab.

Additional e-books in this series can be found on Nova's website under the e-book tab.

INTELLECTUAL PROPERTY IN THE 21ST CENTURY

PATENTS

INTERNATIONAL PROTECTION FOR SMALL BUSINESS AND THE PRIOR USER RIGHTS DEFENSE

SHAWN BENNETT
EDITOR

nova publishers
New York

NOTICE TO THE READER

Library of Congress Cataloging-in-Publication Data

ISBN: 978-1-63483-626-5

Published by Nova Science Publishers, Inc. † New York

CONTENTS

PREFACE

Economic research shows that small businesses are the primary driver of job creation in the United States. Yet, the capacity of American small businesses to create jobs is at risk. American firms compete and grow by supplying products and services that consumers demand, and by internationalizing their businesses through licensing, franchising, or exporting. For many small companies, patent protection prevents competitors from simply copying their innovations, and aids in attracting investor capital needed to grow, build market share, and create jobs. The aim of this book is to show how the U.S. Patent and Trademark Office, in coordination with other Federal departments and agencies, can best support businesses with international patent protection. Moreover, upon its signing in 2011, the Leahy-Smith America Invents Act (AIA) substantially overhauled U.S. patent law. Included in AIA was the expansion of the prior user rights defense to infringement and broadening the classes of patents that are eligible for the new limited prior user rights defense. (Here, prior user rights, is a defense to patent infringement afforded to a party that was commercially using, or engaged in substantial preparations for commercial use of, an invention later patented by another party). In the debate about the impact of an expanded prior user rights defense on the patent system and innovation more generally, the U.S. Patent and Trademark Office produced a report detailing its findings and recommendations on the operation of prior user rights in the industrialized world, which is discussed further in this book.

In: Patent
Editor: Shawn Bennett
ISBN: 978-1-63483-626-5

Chapter 1

INTERNATIONAL PATENT PROTECTIONS FOR SMALL BUSINESSES*

U.S. Patent and Trademark Office

EXECUTIVE SUMMARY

Background

Recent economic research shows that small businesses are the primary driver of job creation in the United States, with young startup companies, which are by their nature small businesses, creating on average three million U.S. jobs per year. Though this pattern of job creation has largely held true for over thirty years, the capacity of American small businesses to create jobs is at risk. American firms compete and grow by supplying products and services that consumers demand, and by internationalizing their businesses through licensing, franchising, or exporting. For many small companies, patent protection prevents competitors from simply copying their innovations, and aids in attracting investor capital needed to grow, build market share, and create jobs. Yet small companies face significant financial challenges in acquiring, maintaining, and enforcing patents outside the United States. Therefore, supporting small firms and fostering job creation requires a

* This is an edited, reformatted and augmented version of a report to Congress issued January 2012.

thorough understanding of these challenges and an exploration of possible remedies.

The Leahy-Smith America Invents Act requires the Director of the United States Patent and Trademark Office, in consultation with the Secretary of Commerce and the Administrator of the Small Business Administration, to study how best to support businesses with international patent protection. The USPTO and SBA are required to determine whether a revolving fund loan program or a grant program would be proper for helping small businesses defray the costs of filing, maintaining, and enforcing international patents. The USPTO and SBA are required to provide other pertinent recommendations.

To prepare this report, the USPTO principally relied upon input from the public, seeking comments through a *Federal Register* notice and holding two public hearings to collect evidence for this study. At each hearing, government panelists from the USPTO and SBA invited and accepted formal testimony, and allowed informal comment and questioning from members of the public. In all, the USPTO received eighteen sets of comments, including responses from a U.S. intellectual property organization, an international industry organization, a foreign government intellectual property office, a U.S. law firm, a U.S. private company, eight U.S. patent practitioners, and five American citizens speaking as individuals. In addition, the USPTO conducted research, relying upon archival sources such as reports, economic studies, and existing data sources where available.

Findings

Available information, including the comments and testimony that the USPTO received, indicates that while patenting appears relatively uncommon among U.S. small businesses, it tends to be concentrated in high technology companies and can aid in securing for them a competitive advantage.

Many small companies grow and create new jobs by following an internationalization strategy, and in this regard international patenting – when done early in the life of a company – can provide a platform for tapping new markets later in life. Evidence also suggests that U.S. small businesses may be patenting less frequently than larger firms, and that they face high costs in pursuing international patent protection. These high patenting costs often occur early in the life of these companies, when funding and cash flows are generally limited. These international patenting costs are also often exacerbated for U.S. small companies because – unlike the USPTO, which gives discounts to eligible small businesses from all over the world – foreign patent offices do not generally provide discounts for small businesses.

Commentary and evidence recognized that other governments around the world are subsidizing patenting by their citizens, with China being the largest and most aggressive actor in this regard. However, public comments reflected considerable skepticism as to whether the U.S. government ought to provide public funding to small businesses for international patenting, instead generally favoring market solutions where possible. Comments agreed overwhelmingly with the proposition that the U.S. Government should engage in diplomacy and patent-system harmonization to help reduce the costs associated with filing foreign patent applications. Moreover, there was general agreement that an aggressive program of small-business education could aid American companies to make informed decisions regarding the optimal international patent strategy. Consistent with the public's reluctance to support government financing, there was no consensus as to whether a revolving loan program or a grant program would be more appropriate.

Recommendations

This report's findings support the notion that many small businesses may benefit from extending patent rights outside the U.S., but too few are aware of the need to do so or the pathways and mechanisms that are available to make these decisions accurately and pursue them cost effectively. To improve awareness and expertise among small businesses, the USPTO and the SBA are positioned to build upon several successful current intellectual property education and training programs. One such program is the USPTO's Intellectual Property Awareness Campaign, focusing on intellectual property basics and offered since 2005 to over a thousand small businesses in various cities throughout the U.S. To reach more small businesses, it may be productive to scale up the current IPAC program through existing partnerships between the USPTO and SBA. The USPTO also recommends ongoing industry engagement to investigate useful approaches to solving the issues raised in this report, including possibly public-private partnerships or other means of helping small businesses.

Public commentary also supports the notion that the public is uncertain about whether the U.S. government should use taxpayer dollars to directly subsidize small business foreign patenting in place of market solutions. There is too little evidence upon which to base sound policy in this area: Neither the academic research nor public comment offered sufficient evidence to determine relative advantages, if any, of the U.S. government employing a loan versus a grant program to help defray the expenses of small businesses seeking international patent protection. Given the lack of data, USPTO does not recommend a program of taxpayer-funded financial assistance to support small business foreign patenting at this time. However, it would be useful and informative as a next step for the USPTO, the SBA, and other allied agencies to collect more information.

I. INTRODUCTION

While firms of all sizes create jobs, recent research shows that small businesses are the primary driver of job creation in the United States. In fact, a relatively small number of American businesses create a disproportionately large share of new jobs. One recent study finds that fewer than five percent of U.S. companies may create more than two-thirds of American jobs, and that these companies on average employ only 61 workers in a given year.[1] Other economic research suggests that young startup companies, which are by their nature small businesses, create an average of 3 million jobs per year, far more than their larger counterparts. This pattern has held generally in the U.S. for more than three decades.[2]

Disturbingly, current economic research shows that these contributions by America's small businesses are at risk. One study found that U.S. small businesses are beginning to generate fewer jobs than would be expected from the historical trend.[3] This study finds that the nature of new, generally small, businesses is changing and moving in the direction of providing less employment. This study also finds that the trend predates the recession that began in 2007. Thus, although small firms remain substantial job creators, their contributions cannot be taken for granted.

American businesses compete and grow in marketplaces by supplying products and services that consumers demand, and by internationalizing their businesses through licensing, franchising, or exporting. For innovators in the American economy, patent protection is often necessary to prevent copying and help in attracting investor capital, thereby allowing these companies to make the necessary investments to grow, build market share, and create jobs. For U.S. small businesses trying to compete in global markets, securing patent protection overseas can be a critical precondition to successfully internationalizing and developing into the productivity powerhouses of tomorrow. Yet small businesses also face significant challenges—particularly financial challenges—in acquiring, maintaining, and enforcing patents abroad. Therefore, supporting small businesses and fostering job creation requires understanding these challenges and exploring possible remedies.

The Leahy-Smith America Invents Act ("AIA") requires the Director of the United States Patent and Trademark Office ("USPTO" or "Office"), in consultation with the Secretary of Commerce and the Administrator of the Small Business Administration ("SBA"), to study how the USPTO, in coordination with other Federal departments and agencies, can best support businesses with international patent protection. The USPTO is directed to make a determination as to whether a revolving fund loan program or a grant program should be established to help small businesses defray the costs of filing, maintaining, and enforcing international patents. The USPTO may also provide to Congress any other pertinent legislative recommendations.[4]

II. BACKGROUND

A. Methodology

In order to obtain statistically rigorous evidence regarding the questions posed by Congress, the USPTO attempted to design and conduct a primary survey of small businesses about their patenting practices. However, the Office was required to complete the study in four months from enactment of the AIA and was permitted to use only its existing resources. These requirements foreclosed the ability to conduct a primary survey of small businesses inquiring about the challenges in patenting outside the U.S. To overcome a lack of information about the extent of the problem, staff in the USPTO Office of Chief Economist analyzed data gathered from the National Science Foundation's 2008 Business R&D and Innovation Survey ("BRDIS"), conducted in early 2009 and housed at the U.S. Census Bureau. Analysis of these data did not, however, support the reporting of any robust inferences regarding the importance of foreign patenting to small businesses or the relationship of foreign patenting to several economic performance measures of interest. Still, the Office was able to gather a number of studies and reports that examined issues related to the questions raised by Congress, but found only a limited body of literature directly examining the questions posed in the legislation. The pertinent findings of this existing literature are discussed below.

In order to report on the key issues, the Office principally sought and relied upon input from the public. Specifically, the Office published a Federal Register notice seeking comments and announcing two public hearings for this study.[5] The Office also provided the public with a dedicated e-mail address and a contact person in the USPTO Office of Chief Economist to receive comments and answer questions. As announced in a Federal Register publication, the Office held two public hearings, one at the USPTO headquarters in Alexandria, Virginia, on Thursday, October 27, 2011, and another at the University of Southern California Gould School of Law in Los Angeles, California, on Tuesday, November 1, 2011. At both hearings, witnesses provided testimony and exchanged comments with the audience in person as well as via teleconference. Representatives from both the USPTO and SBA attended the hearings and actively questioned witnesses. At each hearing, government panelists from the USPTO and SBA also invited and accepted spontaneous formal testimony, and allowed informal commenting and questioning from members of the public.

A. Responsiveness

Through the *Federal Register* notice and hearings, the Office received eighteen sets of comments and testimony.[6] Respondents and witnesses included a U.S. intellectual property organization, an international industry organization, a foreign government intellectual property office, a U.S. law firm, a U.S. private company, eight U.S. patent practitioners, and five American citizens speaking as individuals.

III. Discussion

A. Findings

The academic literature, comments, and testimony that the Office received and reviewed indicate that patenting activity is relatively uncommon among small businesses and, where it does occur, is concentrated in high technology sectors. There is evidence in the literature as well as agreement among respondents that patenting and internationalization can be significant drivers of competitiveness and growth, and boost the job creation potential of a small business. Indeed, patent protection was identified as a key factor to many successful entries into global markets. Yet despite these rewards from international patent protection, American small businesses may be patenting abroad less frequently than larger, more established firms.

Public comments suggest that the reasons why small businesses may be patenting abroad less frequently are largely economic and driven by liquidity constraints faced by many small businesses early in life. International patenting costs can run in the hundreds of thousands of dollars,[7] a substantial expense given the limited resources of many small businesses and start-up firms. Moreover, patenting expenses often occur early in the life of small businesses. At these early stages of innovation and development, such high costs are also likely the most difficult to fund. The result is an often intractable choice for some small firms and independent inventors: to succeed, they can neither ignore international patent protection nor afford it.

Against this problem, the approach of the United States is mostly different from that of other countries. The United States, by providing discounted filing and maintenance fees for small businesses and independent inventors regardless of national origin or industry, takes an impartial, technology-neutral approach that supports innovation by the smallest economic players. By contrast, most other countries do not offer financial support to small entities through discounted filing and maintenance fees, and those that do often use direct subsidies and other government aid that singles out particular firms and industries for support rather than impartial, technology-neutral support. Many such aid programs restrict the application of these benefits to their domestic applicants, to the detriment of American small businesses that seek to compete globally.

1. Patenting by U.S. Small Businesses Is Relatively Uncommon, and Concentrated in High Technology Sectors

Unfortunately, the United States does not have a full picture of how, and under what circumstances, small businesses are using the patent system, whether domestically or internationally. The USPTO has recorded huge increases in the numbers of patent applications being filed at the Office, a trend that reflects both the increasing innovativeness of society and also the value of intellectual property protection in a global economy more and more defined by the production of intangible assets. U.S.-origin patent applications submitted to the USPTO grew from 177,511 in 2001 to 241,977 in 2010, a 36% increase over the decade.[8] Of the patent applications filed at the Agency between 2007 and 2010, about 30% paid small entity filing fees (a discount available to individuals or businesses with fewer than 500 employees), even while by one measure more than 50% of all U.S. utility patent applications in both 2009 and 2010 originated from outside the United States.[9]

Although small businesses have originated approximately one-third of USPTO patent applications in recent years, too little data exists on how, why, or what share of U.S. small businesses are participating in the patent system. Some of the best evidence comes from a large scale survey—not of small businesses *per se*, but of startup businesses in their early years—conducted by researchers at the Ewing Marion Kauffman Foundation.[10] Using the most recent set of these data available, a picture of adjustments suggested by the Kauffman research team,[11] it is possible to develop a statistical representation of patenting by a cohort of U.S. small businesses founded in 2004 in their sixth year of operations.

These statistics show that holding patents by young startups in their sixth year of life is relatively uncommon. Analysis of the data demonstrates that only approximately 2.5 percent of these small-business respondents report any patenting activity, which can include filing applications, being granted patents, or purchasing patents from others. Among industries identified as "high technology," the share of companies with patenting activity is higher at approximately 8.0 percent, although only about 5.5 percent of all small businesses in the sample fall into the "high technology" category.[12]

2. Patenting Can Be Important to the Competitiveness of Small Businesses

While these figures suggest that patenting is relatively uncommon among young small businesses, other evidence suggests that patenting is nevertheless important to high technology companies, and can be associated with superior economic performance. Because the initial cost associated with patent protection is relatively high for early startups, and because patents can have uncertain value, understanding the patenting-success relationship is relevant. Although academic economic research has not answered this question definitively, studies in both the United Kingdom and the United States have found that increased patenting is related to increased growth along several economically relevant dimensions.[13] Similarly, in a survey of young high-technology companies conducted in 2008, a group of researchers at the University of California, Berkeley found that patenting is quite common among biotechnology and medical device startups. When these companies had venture-capital funding, the share of companies engaged in patenting exceeded 90%.[14] Moreover, patenting was not uncommon even among information technology startups, including both software and internet companies.

Many high-tech startup executives stated that patenting was important for capturing competitive advantage in the marketplace, preventing copying, improving success at attracting investment, and increasing the likelihood of being acquired by another company or having a successful initial public offering (IPO). These last findings are supported by other economic researchers who have suggested a positive relationship between startup patenting, meaningful early investor funding, and successful transitions into larger publicly-traded firms.[15]

3. Internationalization Strategies Can Facilitate Small Business Growth and Job Creation

In an increasingly global economy, internationalization strategies can be effective mechanisms to access markets, serve unmet demand, and grow small companies, thereby increasing manufacturing, production, and job creation. Various ways of entering non-domestic markets—such as licensing, franchising, exporting, and foreign direct investment—have been shown to be related to the growth and successful performance of small companies.[16] In addition, the U.S. government has recognized the importance of exporting to the economic health of the nation, and has taken action in the form of programs like the National Export Initiative (NEI).[17] The NEI also explicitly focuses on the importance of supporting exports by small and medium-sized enterprises.[18]

4. Patent Protection Abroad Opens Opportunities for Successful Entry into Global Markets

Economists recognize that a well-developed system of property rights, particularly for innovative companies, is often a precondition to successful internationalization.[19] The comments received in response to this study reflect a similar understanding among the IP and business community. For example, one respondent noted the importance for small- and medium-sized enterprises (SMEs):

> International patent protection is critical for SMEs that have inventions that are viable both in the domestic market and overseas. The international marketplace is extremely competitive as it is. American SMEs need every possible edge over their foreign competitors to secure on-going sales. Having patent protection for an advanced technology, or invention isn't a guarantee, but is a significant advantage.

From an economic perspective, patent rights can be thought of as options. While one must pursue patent protection today, the protection can have an extended life and thus provide future opportunities to engage in later productive economic activity.[20] One witness recognized this reality:

> A failure of SMEs with international strategies to obtain international patent protection early dramatically increases the risk that another business will beat them to the patent office and effectively block any possibility of growth in that country.[21]

These considerations also have technology- and industry-specific implications. One respondent from the electronic and software industry observed:

> The key reason [sic] that a small business needs patent protection outside the U.S. are those cases in which they wish to license another company to make and sell the product in countries outside the U.S. with the expectation that the product will never enter the U.S.[22]

Yet many small business owners are unaware that patent rights are territorial, and the protection offered by a U.S. patent ends at our borders.[23] Protection in other markets must be sought on a nation-bynation basis. And many U.S. small businesses do not realize all the paths that they have at their disposal. One witness pointed to the lack of information among small businesses about acceleration mechanisms in international patent prosecution, such as work-sharing among patent offices and the Patent Prosecution Highway.[24]

In order for a U.S. small business to maintain an option of selling or making its innovation in another country, it is often necessary to patent the innovation abroad. By doing so, a U.S. small business can prevent others from patenting the invention in those nations and prevent foreign competitors from simply copying the invention. Moreover, academic research has suggested that small business success in internationalization can be aided by partnering with larger companies with resources and expertise.[25] That may include collaborating with an established partner in the foreign market, which can require the possession of foreign patents. According to one witness:

> International partners will want to see local protection for their markets if they want to collaborate with the small company or invest/purchase. It is critical.[26]

The comments also noted the concentration of patenting in high-technology firms and discussed the relative scarcity in fact of patent-holding by young high technology firms. As to when small businesses should take an active interest in securing international patent protection, one respondent noted as follows:

> At the outset, international patent protection may be critical in the first few years of establishment. For startup high-tech businesses that have invented the core idea of a new technology, it is critical.[27]

Yet as to when small businesses actually do take an active interest in securing international patent protection, the same respondent also noted as follows:

> The small enterprise may realize that international patent protection is important, but it may not include that component in its business strategy until it is too late. Competitors in other countries may already have copied and used the invention, and even patented the same or similar inventions or their improvements, to the detriment of the original inventor.[28]

The comments in this regard were not uniform, however. Another respondent from the industrial equipment sector discussed the importance of international patenting not in terms of a company's high-technology focus, but rather of general marketability and foreign demand:

> International patent protection is crucial for SMEs whose inventions have [broad-based] appeal, are competitive and therefore have excellent export potential, versus those inventions that are limited in scope, or cannot compete on a cost [basis] with similar foreign inventions, and thus would be economically viable only in the U.S. domestic market.[29]

Other respondents suggested different reasons for international patenting, both those related to the reliability of enforcement and the availability of established partner firms in the pertinent foreign jurisdiction[30] and benefits associated with building markets and attracting investors.[31] Not least, one respondent specifically highlighted the absence of detailed empirical studies into what characteristics determine the importance of international patent protection for small businesses in different technology fields and in national markets of different size and distribution.[32]

5. Small Businesses May Be Obtaining Patent Protection Abroad Less Frequently Than Large Companies

In 1995, the U.S. Small Business Administration (SBA) commissioned a statistical study that found U.S. small businesses possessing U.S. patents issued in 1988 were less likely than their large counterparts to extend their patent protection to other countries.[33] This study used metrics suggesting that the difference could not be explained by small businesses producing less valuable or less important inventions, or engaging in technologies where patenting or foreign business was considered less important. Nevertheless, though the SBA report suggested that the lower rate of foreign patenting among small businesses was not driven by lower-value innovation or industry effects, the report's findings were not conclusive.

A follow-on report commissioned by the SBA and released in 2003 used statistical analysis to examine U.S. patents granted to U.S. small businesses during 1988, 1992, 1996, and 1998.[34] By comparing these patents to those held by large companies, the report found that small businesses were extending their patent protection outside the U.S. less frequently than large entities irrespective of technology field, suggesting that the differences were not a product of these companies working in different industry sectors.

Neither of these SBA reports, however, discussed whether the differences in foreign patenting behavior had any relationship to the success of these companies, or their performance in terms of growth, employment, sales, or profits. Also, there was no evidence presented that larger companies extended protection abroad at the optimal level, or that small businesses, by comparison, extended protection to a sub-optimal degree.

Commentary at the hearings discussed these issues. One respondent spoke of the competing interests in the early life of the company, and why international patenting may happen less frequently at small and medium-sized enterprises (SMEs):

> SMEs also tend to see foreign patents as a low priority benefit in the early stages of the business. Priorities are finalizing product design, marketing and sales to result in "winning" in the marketplace, and obtaining a U.S. patent position if appropriate.[35]

Indeed, another of the comments suggests that this difference is driven largely by the high cost of obtaining patent protection internationally. According to one witness:

> International patents are expensive, and small businesses have other competing priorities for their capital. . . . With regard to the acquisition of international patent rights, translation costs, annuity fees, and foreign professional fees represent significant expenses and act as barriers that often prevent small businesses from applying for foreign patents.[36]

The problem of high cost in international patenting is further complicated by the limited time in which to decide where to pursue national stage patent protection spun off from a Patent Cooperation Treaty application:

> For example, a biotech company that files a U.S. patent application today (and a PCT application one year from now) has only 30 months to decide whether to abandon the application if it wants to avoid the cost of entering the national stage in a number of foreign countries. 30 months may be enough in some other industries, but in biotech that's too soon for an informed decision. Including translation costs, the aggregate expense of entering the national stage in Japan, Korea, Europe, Australia, and the NAFTA countries can easily exceed $100,000; if the BRIC countries are added, costs can double.[37]

Moreover, U.S. small businesses perceive working capital to be even more important in light of the firstto-file provision of the AIA:

> And this need for cash comes even earlier under the America Invents Act. Since the U.S. will be a first to file country, U.S. filings will come as soon as possible and the foreign filing decisions will be stepped up. Moving expenses up for an early stage small life science company is more difficult

and the probability increases that these early inventions are not adequately protected.

Small companies usually do not have much available cash which makes paying foreign filing expenses more difficult. They run leanly and need to periodically raise capital to fund their operations. However, they do not raise more money than necessary in the short term it would require selling too much equity at a low price.[38]

Nor does the high importance of international patent protection appear to depend on technology field, according to one respondent:

International protection is critical for any small business that develops a new drug or medical device, since there may be no other clinically acceptable alternative. A patent on that new drug or device will ensure exclusivity for an extended period of the drug or device's market lifetime. In the wireless and IT sectors, if a small business does not have patents in as many countries as possible and as soon as possible, it may be very difficult to obtain local investment in its venture.[39]

Indeed, the relative importance of international patent protection may, to an extent, be altogether unknowable for many firms early in their development. Patenting abroad becomes significant, according to one respondent,

usually 3-6 years after the invention has been made at the earliest, more often, 10 years after the invention has been made. When a small business is starting, they cannot predict very with much accuracy where the market for the product will develop, who the primary buyers will be across the globe and the time it will take to move a product from the idea stage to the market stage.[40]

Moreover, most small firms plan to patent internationally based not on their own manufacturing or marketing operations, but rather on an established, licensed partner firm:

In the context of a small business, they do not expect to be selling product in many countries; if they did, they would be a large company. Nearly all small businesses do not have the production and market distribution capability to sell internationally themselves. Rather, they will

> achieve this by licensing an established company who has such . . . distribution chains available. The other company might be a large U.S. company, a large company in another country or a very large multinational company.[41]

The USPTO public hearing in October revealed an illustrative pair of case studies on the importance of foreign patenting to the success of a small business. The first firm dealt in traffic surveillance technology. The second firm dealt in imaging technology.

a. Case Study: Traffic Surveillance Technology

In studying Washington, DC, traffic patterns, a group of professors at the University of Maryland recognized that most modern car keys employ short-range wireless transmitter technology such as Bluetooth®.[42] By detecting each key's wireless identifier as cars pass by, it is possible to determine how many cars are on the road and how fast each is going, all in real time without the need for cameras. Extensive research and development went into using this technology to manage traffic through existing infrastructure.

The professors then turned to business strategy. They obtained patents on their invention and formed a company with support from the university to commercialize their invention. They obtained venture capital funding and invested in competent personnel. And they sought clients, expecting first to establish their brand in the greater Washington, DC, area, which is known for its severe traffic congestion problems. Yet their first business came in foreign markets, including Singapore and Sweden.

The firm took a far-sighted approach and pursued international patent protection, including the substantial cost of foreign translation. The firm did so at a time when it had limited capital. Because the firm invested in protecting its promising technology in multiple foreign markets, it has now built an internationally successful brand that cannot easily be copied by foreign rivals. And where the firm wishes to deploy its traffic surveillance technology through other companies who can better meet local demands, it is strongly positioned to capture licensing revenues.

b. Case Study: Scanning and Imaging Technology

A small American imaging firm operated in the market for scanners.[43] The owner had previously been an employee at Kodak and had a sophisticated understanding of imaging technology. The firm's scanner products commanded quite high prices in the market, upwards of $10,000 per unit, as the scanners themselves reflected the state of the art. Not surprisingly, a great deal of patented technology and software went into developing and producing the firm's scanners.

The firm's owner, however, met with its intellectual property counsel and learned that acquiring international patent protection, particularly through the PCT process and in light of the translations required for multiple foreign jurisdictions, was procedurally complex and quite expensive. Therefore, management made the business decision to obtain patents only in the United States and in Australia, but not, for example, in Europe, Japan, or China—major trading partners of the United States.

Soon after, at an industry conference in Germany, the owner discovered that a German firm was demonstrating scanner products that used the very same core technology that he had pioneered. The German company was well-funded and successful, and was quite candid that it had derived its products from the American inventor and his small business. Yet the German firm highlighted its considerable corporate backing and network of clients and existing service agreements, and carried away a great deal of business as a result.

The American inventor consulted his intellectual property counsel regarding the situation in Germany, but learned that he was bound by his earlier economic decision to abdicate patent protection in most foreign countries. Unless the German firm imported its scanners into the U.S., disputing the expropriation was, ironically, not worth the option.

Later, testifying before Congress on this issue, the American small business owner lamented that his intention had never been to drive the German firm out of the market. To the contrary, he would have preferred to do business with them: grant them a license and mutually enjoy the benefit of their resources and distribution networks. Only his early and difficult economic choice left him with no bargaining position. He had unwittingly made a royalty-free donation to an overseas competitor only too grateful to capture the market and the jobs, without taking the risk or expending resources on research and development.

6. International Patenting Costs Are Often Substantial

In comparison to the United States, patenting in other international markets tends to be expensive. It may be impossible to compare the entirety of economic costs and benefits of patenting in one nation versus another due to differences in patent scope, patent laws, and patent enforcement. Nevertheless, it is possible at least to compare the levels of patent-office fees faced by a typical small entity seeking patent protection in different countries.

Significantly, the U.S. recognizes applicants who qualify for "small entity" status and provides a 50% fee discount to such entities and, under new law, a 75% discount to a "micro entity."[44] By comparison, those patent offices around the world which U.S. companies principally target for international protection tend to charge much higher fees.[45] Foreign patent offices also tend to be substantially more expensive for U.S. small businesses because of the undiscounted annual maintenance fees charged.[46]

While there are substantial differences in the fees paid to patent offices to secure and maintain a patent, such fees only represent a fraction of the total costs of patenting paid by small businesses. Small businesses pay patent attorneys for their services and for required language translations. Translation is expensive in general, and technical translation significantly more so. Recent economic research suggests that required translations into the national languages needed in many international patent offices can be a significant disincentive for companies to seek patent protection in these other countries.[47] Including translation costs, the General Accounting Office in 2003 estimated a cost of between $160,000-320,000 in foreign patenting costs to a U.S. company seeking protection in four foreign jurisdictions (Japan, Canada, South Korea, and the European Patent Office). Such costs will necessarily depend on which countries are chosen and the complexity of the invention.

The IP community has specifically noted the disparity. One witness discussed it in terms of the broader problem of cost to small businesses. A first cost is official fees:

> The overall official fees for filing are higher in most foreign jurisdictions than in the US, and there are no small entity discounts. Excess claims fees are particularly high in the EPO.[48]

A second is translation costs:

> The cost of the lawyer is actually less than the cost of the translation in certain cases. So what you would think, the highly trained, really valuable patent lawyer guy that says, that's going to cost me a fortune, but when I add up my translation cost, that actually flips the equation, I am spending more to get the translation than I am on the attorney.[49]

A third is the cost of outside counsel, both in the U.S. and abroad.

> SME discounts are rare overseas, if available at all. Fixed costs and hourly rates for attorneys in, say, London or Tokyo are perceived to be significantly higher than those of their U.S. counterparts. Foreign firms do not have the kind of SME-serving culture that exists in, say, the California bar.[50]

Furthermore, in comparison to the United States, the process to secure a foreign patent can be overwhelming. A separate application is generally required in each country in which protection is sought. There are two international procedures that can be useful in obtaining protection. One, established under the Paris Convention, allows an ordinary application to be filed in each foreign country in which protection is desired, within one year of when the application is filed in the United States (or other first country). The second method, the Patent Cooperation Treaty (PCT), must also be pursued within a year of original filing; it allows for a form of international patent search to be carried out, but there remains substantive and administrative processing needed for each treaty-member country in which protection is desired, plus the payment of fees and other costs. One advantage to using the PCT route for the applicant is that it allows one to defer these various "national-phase" examinations for at least 30 months after the original filing.

Comments reflected that the PCT method can be a useful one for a small business, with one witness noting that the PCT "is the preeminent vehicle for pursuing international patent protection."[51] The USPTO also received testimony from a practitioner whose twenty years of experience has included "substantial use of PCT practice."[52] The witness explained the benefit of using the PCT system:

> I can count on ... one hand the number of times where I've gone direct national rather than using the PCT system. So I think qualitatively it's done a very good job to facilitate the winnowing out of cases that are weaker than others.[53]

Yet the same witness also qualified his regard for the PCT:

> But I think the critical limitation with respect to the PCT system in general is that you have to file 12 months before you have all the information that you garner from the PCT process. And so if you have information available to you at the time of foreign filing or entering the process, you will be able to make much better business decisions about those patents that you would proceed with.[54]

Beyond procedural limitations, moreover, some comments about the PCT also identified drawbacks to using the system. One respondent framed the issue as one of fee deferral:

> [Most startups] also view PCT solely as a fee deferral system, mainly because the International Search Reports prepared for their technologies do not result in useful information, and because examining standards for IT-related subject matter differ greatly around the world, making centralized amendments is not particularly useful.[55]

7. Patenting Expenses Often Occur Early in the Life of Small Businesses and Are Difficult to Fund

Because small businesses are often cash-constrained, the amount of capital and funding they possess can be an important limitation on their ability to obtain patent protection. An early-stage startup may be pursuing patent applications, both in the U.S. and abroad, at the same time as it is seeking and obtaining funding and making other early-stage investments in equipment, development, and hiring employees. Often, the point at which capital is necessary for investing in patent protection—which must be expended early in the development of a new innovation due to patent priority rules—will require that the company forego other investments in its future growth and success. This suggests that uncertainty about technical and commercial potential of the business plays a huge role, and the relationship between the initial funding cycle for young companies and the initial patent lifecycle is an important consideration.

While small business owners often use personal funds, credit, and cash flows from operations to fund growth, many also seek funding from external sources. They may obtain loans from commercial lenders or equity investments from friends and family, so-called "angel" investors, or venture-capital firms, among other sources. Many startup companies fail, and the survivors are often cash-constrained for a number of years in their early lives.[56] A small number of startups will develop into "gazelles," fast growing companies between 3 and 5 years of age. A recent economic study suggests that these "gazelles" account for less than 1 percent of all U.S. companies yet generate about 10% of new jobs per year.[57] This same research finds that throughout the American economy in recent years, just 1% of companies, the top performers, contributed as much as 40% of the total employment growth, and these top performers are heavily concentrated in companies of less than 500 employees.[58]

The extension of patent protection during the last several decades to include software and business methods has increased the pool of U.S. small businesses that may seek patent protection for their innovations. When an innovating company seeks patent protection, it faces a sequence of costs that must be paid to patent offices (in fees) and to patent practitioners (for preparing documents) at a time when the typical startup company is cash-constrained. Thus, small businesses may select against patenting early in life and face later, negative ramifications. According to one witness familiar with the information technology (IT) industry:

> Venture capitalists and other early-stage investors in IT businesses tend to view patent exclusivity as a secondary factor because the real problem is competing in the market against established behemoths on the merits of product features and functions. Patents become more important in years 3 and later when the future of the business is more apparent, second-tier investors have entered or a revenue stream exists.[59]

On top of what may be large expenditures for research and development by the small company, the total cost to obtain even U.S. patent protection can average almost $40 thousand dollars,[60] and increase considerably depending on the complexity of the invention and on prosecution options such as appeals and interferences.[61] These costs may be multiplied many times over if the small business seeks to extend its patent protection in other nations.[62]

The timing of international patenting costs is important when compared to the funding timeline of the typical early-stage company. Up-front patenting costs are significant, with a large portion coming in the form of attorney costs and translation fees. At the earliest point in a young company's life, cash flows tend to be the most limited and constrained. Often, early-stage companies have negative cash-flow, so any investment in patenting must either come from personal funds, debt, investor capital, or be foregone altogether.

In addition, the returns to patenting may be uncertain. A substantial body of economic literature shows that the value distribution in patenting is highly skewed, with only a relatively small fraction of patents producing large financial returns to the patentee.[63] Unfortunately, patenting is required to be initiated at the beginning of the innovation process, when uncertainty over the technological potential of the invention, and its commercial potential in the marketplace, tends to be high. Therefore, small businesses must often make patent-protection investments early during a situation of high uncertainty, while not fully able to predict how meaningful the patent protection may be to the company's future competitiveness. Such uncertainty may result in a small business either under-investing or over-investing in patent protection in these early stages.

Under normal circumstances, information about a company's prospects should become apparent over time as the business tests market demand. When markets for entrepreneurial capital are operating reasonably efficiently, investors should be willing to put capital into businesses that appear to be good investment candidates. So over time, capital should be available to these companies so long as they are better able to demonstrate the commercial potential of their inventions. Under such circumstances, successful small businesses ought to be better able to bear costs that occur as the company grows and later in the patenting life cycle. Because later patenting expenditures, such as renewal fees and enforcement costs, generally occur many years after the initial patent application filing phase, there may be less reason to believe that the markets for entrepreneurial capital are failing systematically many years after the founding of companies.

Many of the comments and testimony supported this concept. In several instances, witnesses observed that public support for expenditures later in the international patent protection process is not optimal, tending to disfavor later-stage funding assistance on a number of scores. Industry representatives suggested that late-stage funding is normatively inappropriate for activities such as maintenance:

> The start-up business that has found the money to remain in business for five to ten years when most maintenance expenses are first due should use its own money to maintain its patent rights. AIPLA does not believe that there is a role for the USPTO or other federal agencies in such activity.[64]

Another witness posited that late-stage funding is not undersupplied in activities such as enforcement, particularly because by then the patent information tends to be better developed (but not always):

> With regard to enforcing international patent rights, I think much of the focus today has been on the procurement side, and certainly that's where the expenditures are most acute because the technology is not yet proven, so it's an investment.
>
> On the enforcement side, generally speaking, for two parties to have a dispute, there is already [a] value proposition associated with the patent. So I think the issues, in my opinion, at least, are less acute; however, if you get to the stage of enforcement, if you find your patent is not worth what you thought, that clearly is a problem.[65]

Another witness echoed this view of late-stage funding as simply less important than early assistance for filing and prosecuting patents:

> QUESTION: Given limited funds, whether it's a loan program or a grant program or if there are funds available, when are those funds most important? In the front end, filing, or what the legislation also envisions as maintenance? When is it more important?
>
> WITNESS: I think it's unquestionable that it's front end, because if I'm going to be successful, somebody is going to give me money for maintenance.[66]

8. Foreign Countries Largely Take a Different Approach Than the United States in Assisting Small Businesses to Acquire Patent Protection

Since 1982, the U.S. government has chosen to support patenting by small entities by discounting USPTO fees by 50 percent for independent inventors, small businesses, and non-profit organizations through a "small entity" status designation.[67] Congress originally enacted the small entity discount to alleviate the burden of 1982 increases in patent filing, processing, maintenance, and issue fees.[68] At the time, maintenance fees themselves had only recently been instituted, in 1980.[69] Significantly, the fees charged and fee discounts offered by the USPTO apply to both U.S. applicants and foreign applicants.

The motivation for these benefits is an appreciation by Congress that independent inventors, small businesses, and non-profit organizations seeking patent protection for their innovations should not bear the same economic burdens as well-established firms and corporations.[70] Indeed, the AIA goes further, adding a seventy-five percent discount for smaller "micro entities."[71]

The large majority of countries do not accommodate small entities in the same nondiscriminatory manner. Canada and Mexico are notable exceptions. Canada offers a 50 percent discount to entities that employ 50 or fewer employees or are universities,[72] while Mexico offers a 50 percent discount in filing fees to domestic or foreign individuals, independent inventors, small business concerns, and non-profit institutions.[73]

Beyond these, Japan and South Korea provide certain accommodations, but these incentives are not available to foreign applicants on an equal basis with domestic applicants. Japan recognizes small entity status for applicants who are supported according to the Daily Life Security Act; who are Japanese individuals not subject to a residency tax or an income tax; who are companies with capital less than 300m yen, are not subject to a corporation tax, and are not controlled by any other corporation; or who are national universities, national technical colleges, university-cooperative organizations, or small- or medium-sized R&D enterprises. Where the applicant is also the inventor, the individual applicant in the Korean Intellectual Property Office may receive up to a 70% reduction in fees, but otherwise small entities from other countries are not eligible for discounted fees. The European Patent Office and the State Intellectual Property Office of China do not offer any small entity discounts at all.

In contrast to the U.S.-led policy of neutrally applied fee discounts and impartiality as to technology sector or the national origin of applicants, a number of countries have pursued a policy of direct subsidization. China may be the most notable example. In 2006, China's *Outline of National Medium- and Long-Term Science and Technology Development Plan (2006-2020)* announced a new "innovation strategy" to the international community with the goal of "advancing China into the rank of innovative countries by 2020."[74] These initiatives included intent to invest more than 2.5% GDP in R&D, reduce dependence on foreign technology by 30%, and dramatically increase the number of annually granted "indigenous invention" patents.

Since then, China has made concerted efforts to broaden its intellectual property system[75] and increase both domestic and international patent filings. Direct government subsidization of Chinese citizens' patent applications and maintenance fees has been a particularly notable mechanism for supporting this increase. Chinese government subsidization of patenting can occur through two different channels, including (1) direct subsidization by the State Intellectual Property Office of the People's Republic of China (SIPO), and (2) local government subsidization for patent applications and maintenance.[76] The importance of these subsidies to meeting the patent filing goals is clearly spelled out in SIPO's "*Guidance Regulations on subsidization work of patent application*," which asserts that subsidization promotes patenting of innovative ideas, and improves native technological innovation. Additionally, the increased emphasis on subsidized patent support (at lower monetary levels for applications and higher ones for successful grants)[77] can be seen through such examples as:

a) Beijing municipal government's 2006 stated intention of awarding citizens RMB 1,000 for patents applied for in foreign countries ("*Reward Measures on Patent Application in Beijing*"); and
b) Tianjin municipal government's 2006 stated intention of awarding citizen's RMB 5,000 for patents granted in other countries ("*Subsidization Measures of Patent Fees in Tianjin*").

Many provinces have also amplified their subsidization level per patent in recent years. For example, Zhejiang province's local subsidization amount rose from an initial 2,000 RMB per patent to 4,000 RMB in 2006. This may be due to variety of factors, but establishment of patent growth targets within some local governments (e.g. desired 30% annual growth in Anhui province in 2007) likely contributed to augmentation of local subsidization. At the same time, there is no evidence that such subsidies or patenting targets have increased either innovation or economic growth.

Other structural policies also greatly contribute to an increase in patenting. These include such items as mandatory "Inventor Remuneration" (Articles 16) and mandated first filing in China for inventions created there (Article 20) from the Third Amendment to China's Patent Law (2009).[78]

9. In Response to the Questions Presented by Congress, the USPTO Has Identified Points of Agreement Among Respondents and Witnesses as to What the U.S. Government Can Do to Help Small Businesses with International Patent Protection

An overwhelming majority of respondents proposed that the United States work with foreign governments to reduce the costs associated with filing patent applications overseas and to harmonize legal requirements, particularly through existing treaty arrangements.

a. The U.S. Government should engage in diplomacy and harmonization to reduce the costs associated with filing foreign patent applications.

With respect to reducing foreign patenting costs through diplomacy and harmonization, one industry organization reported as follows:

> With regard to the acquisition of international patent rights, translation costs, annuity fees, and foreign professional fees represent significant expenses and act as barriers that often prevent small businesses from applying for foreign patents. The USPTO and Federal Government could work with foreign governments to delay the requirement for submitting a translation, especially in countries with deferred examination. They also could work to reduce annuity payments for small enterprises during the pendency of an application.[79]

A manufacturing firm reported as follows:

> If possible, help negotiate down the cost of filing for small inventors. The initial filing costs for patents are daunting. The fees vary by economy and routinely range from hundreds to thousands of dollars per filing. This does not include the mandatory annual maintenance fees and continuous legal fees related to the prosecution of the application, which could take years to complete. There are a number of major markets that are not part of the PCT. Even with a PCT filing, the application still needs to be filed and prosecuted individually in each member economy. As a result, many SMEs do not file for patent protection in markets outside of the US, because they simply cannot afford to.[80]

With an academic perspective, one practitioner-turned-law professor reported as follows:

> Reduction in filing fees for small businesses for PCT applications would help significantly. In particular, if the PCT application filing fee is lowered to a level equivalent to or less than the typical search fee charged by a search firm for a patentability search, then it is well worthwhile for small businesses to file a PCT because they can obtain patentability search results from the PCT prior art search.
>
> Even if the overall fee associated with a PCT application cannot be reduced, the PTO should consider reducing at least the PCT search fee. It is not uncommon nowadays for patent counsel to suggest to their clients that they go and get their PCT search done in Korea instead of the U.S. because a search in Korea costs about half as much as a U.S. PCT search (about $1000 in Korea compared to about $2000 in the U.S.).81

A manufacturing firm reported as follows:[82]

> There is a complete lack of harmonization in the foreign patent application process. In fact, a patent could be granted in one economy, yet rejected in another. This uncertainty is extremely frustrating for SMEs. And, unlike large companies with full legal departments, SMEs do not have the deep pockets and expertise to navigate through the quagmire of antiquated patent processes in multiple foreign countries.
>
> Ironically, this problem disproportionately affects SMEs in the information technology, software, internet, and social media sectors—the new frontier and high growth segments that support many of the best paying jobs.

These respondents reflect a general consensus that procedural reform, not direct subsidization by the U.S. taxpayers, is the principal remedy for the challenges faced by small businesses in obtaining international patent protection.

b. The U.S. Government should approach the direct subsidizing of foreign patenting costs with care

The comments were split about, but mostly disapproved of, certain kinds of public financial assistance, including loans and grants, from the U.S. government to private industry, but expressed receptiveness to other forms of financial assistance. One industry organization opposed government funding but supported tax incentives:[83]

> The Federal Government should [not] become involved with subsidizing the filing, maintaining, and enforcing of patents abroad. These are private sector issues that are best addressed by businesses, whether large or small. Nonetheless, it may be useful to study the benefit to small enterprises and to the nation of tax breaks for the cost of obtaining patent protection abroad. Also of interest would be profits made through licensing foreign patent rights abroad and the sale of products or services abroad that are covered by such patent rights. In a similar vein, the U.S. Government may consider strategies to encourage sales abroad, for example, by allowing patent costs to be expensed rather than capitalizing them.

A legal practitioner with nearly 30 years of experience working with hundreds of businesses, mostly small companies, opposed government loans categorically and disfavored other forms of assistance:[84]

> I do not endorse a government loan program to help businesses get foreign patents. I oppose government loan programs generally, as I believe the private sector does a better job at evaluating the risks of a loan.

By comparison, a legal practitioner in information technology with over 22 years of experience distinguished loans and grants funded by tax revenues and those funded by user fees, and opposed both:[85]

> A government loan or grant, funded from taxpayer dollars or a portion of existing USPTO user fees, for the purpose of foreign patent procurement by SMEs, does not represent good policy. (a) It would represent, to at least some extent, a redistribution of U.S. income into the social welfare programs of foreign countries with only speculative benefit to SMEs due to the subjective nature of obtaining patents. . . . (b) Further, a material percentage of SMEs either fail or have limited-vision management: C-level executives who are incapable of growing the company or pursuing products with a limited or immature market. It may be difficult for government to determine which SMEs merit a loan or grant. (c) Finally, it seems fair to place at least some of the burden of financing SME patents on the large entities that dominate the system and, because of their ability to pay high official fees, file large numbers of cases, and wait out the resulting backlog periods involved in foreign patenting, must be seen to contribute to the costs faced by SMEs in the system. For all these reasons, I disfavor a government loan or grant program.

However, opinion was not one-sided. An industry organization invited government support building on existing small business programs:[86]

> For our small businesses, securing IP protection is as important as obtaining laboratory equipment, leasing space, or hiring creative, dedicated employees. And because IP business assets are at least as important as other, more tangible business assets, there is no reason to exempt patent rights from publicly-funded small business assistance programs that are available for more tangible assets such as capital equipment, hiring, or leasing space. Extending the range of public assistance programs to patent rights for small businesses would help small biotechs spend money normally allocated to patent filing and prosecution elsewhere.

These comments reflect the IP community's diverse positions on government support for foreign patenting by small businesses. Even those who oppose government aid oppose it to different extents and for different reasons. Nevertheless, the comments do also reflect a point of consensus: a need for greater education for small businesses on IP in general and foreign patenting in particular.

c. The U.S. Government should pursue an aggressive program of education for small businesses on foreign patenting.

Many respondents and witnesses argued that what small businesses, start-up firms, and independent inventors want from the U.S. government is not capital. Instead, to facilitate more informed acquisition of international patent protection, respondents suggest that greater education about the importance of international patents and ways to obtain them is necessary. One industry organization reported as follows:[87]

> In another area, the USPTO and other Federal agencies can do a much better job educating small businesses about the importance of international patents and strategies for effectively pursuing international protection.

A multinational U.S. law firm with a significant intellectual property practice reported as follows:

We believe that the key to spurring competitiveness in small business is education about the role international patent protection can play. Rather than fund particular industries/companies, we believe that the U.S. Patent and Trademark Office (PTO) should continue to provide educational workshops/seminars in addition to information on its website. Those should be held through regional small business organizations in order to provide the most direct impact.

10. In Continued Response to the Questions Presented by Congress, the USPTO Offers Some Observations Regarding Loan Programs and Grant Programs

Consistent with public commentary, the USPTO does not recommend using public funding to directly support – through either loans or grants – small business international patenting at this time. To be responsive to Congress, however, and as background, it is noteworthy that whether a patent subsidy program is structured as a grant or a loan has implications for how effective it will be, and for what kinds of recipients it reaches. A grant—provided it is limited to supporting expenses associated with filing small-business international patent applications—would provide immediate capital to be used only for obtaining foreign patent protection. Because such capital would be available with no repayment obligation, any eligible small business that is interested could take advantage, so long as it had the resources available to obtain and administer the grant internally. As a matter of basic economics, if the grant covers 100 percent of the expenses associated with international patenting, an over investment in such activity might occur.

In order to combat this possibility, grants are commonly offered with requirements that the awardee invest company capital in some kind of matching scheme, thus investing its own capital in some percentage to the grant capital. This percentage need not be equal shares. So, a grant may be designed to cover only some fraction of the total cost of an activity, and in that way, encourage investment by the business without supporting frivolous or trivial expenditures. Such a grant scheme would tend to support an activity that the government has concluded is desirable, while avoiding the harshest outcomes of government money being used to support activities that have nearly zero private value. Such a matching requirement, however, may affect small businesses disproportionately since—as was discussed previously—small and particularly young companies may not have sufficient cash flow or capital to take advantage of this type of grant scheme in the early years of patenting when such expenditure is necessary.

A loan program can also provide up-front funding, but in contrast to a grant, comes with an obligation to repay the capital later, ordinarily with some interest rate. Loan programs can differ across several dimensions, such as how much capital is available, how long a repayment horizon the recipient will be allowed, and in both the rate(s) of interest and the interval(s) at which interest payments will be due. A government-supported loan program is often economically equivalent to a combination of an ordinary, private loan with an economic subsidy or grant. The subsidy represents the extent to which the loan is more attractive than privately available options, such as offering a means of borrowing more money than otherwise possible, or being allowed a more attractive interest rate or repayment date. In this way, a government-supported loan program can share the attributes of a grant program with a matching requirement. Each presents some kind of subsidy to the small business, but also requires some liability. The main difference comes when that liability is realized. In a grant program, it is realized immediately, while in a loan program, it will be realized after some delay. Also, since loans present the possibility of default if the business fails or is otherwise unable to pay, that liability may not be realized by the small business at all. Since loans consist in part of the borrower's own future assets, the borrowers will be less likely to use them to acquire low-value investments. Thus, loans also present a solution, like matching, to the problem presented by a 100 percent grant.

The question of what type of program is better intersects with problems of uncertainty characteristic of the patenting and innovation process described earlier in this report. A patent is generally filed early in the innovation process when uncertainty is likely high concerning both technological and market potential. Without full information, a small business can underinvest in activities which would be valuable, or overinvest in those which turn out to offer poor returns.

Economic scholars appear to have not devoted significant attention to the value of grants compared to loans to help small businesses with international patenting. Although evidence-driven studies would be a preferred way to compare the policies, a literature review found no analysis that directly compared the effectiveness of loans and grants for a program of the kind described here. So, while the USPTO does not advocate using federal funds for these purposes currently, it must also be recognized that we lack sufficient data to adequately understand whether grants, or loans, or neither mechanism, would be effective.

B. Recommendations

1. The U.S. Government Should Engage in Diplomacy and Harmonization to Reduce the Costs Associated with Filing Foreign Patent Applications

Consistent with public comments received, USPTO recommends that the United States continue to engage our foreign trading partners in efforts to reduce patenting costs faced by American small-businesses abroad. The lack of small-entity discounts in most of the world's patent offices discourages U.S. small businesses from patenting abroad while at the same time USPTO's small entity discounts encourage foreign small businesses to file in the U.S. Moreover, efforts made toward substantive harmonization of patent laws around the world can aid U.S. small businesses by allowing them to economize on the costs of filing and fulfilling requirements in different offices around the globe.

2. The USPTO and SBA Should Partner in an Expanded IP Education and Training Initiative Aimed at American Small Businesses

This report's findings support the notion that many small businesses may benefit from extending patent rights outside the U.S., but too few are aware of the need to do so, or the pathways and mechanisms that are available to make these decisions accurately and pursue them cost effectively. Consistent with commentary received from the public, many small businesses would benefit from training in how IP protection and enforcement can be relevant to a growing company in a global economy. Piracy, counterfeiting, and the theft of intellectual property pose a serious threat to many American businesses, whether operating inside or outside the United States.

The USPTO and the SBA are positioned to build upon several successful current IP education and training programs aimed at small businesses, and increase the availability, penetration, and participation in these programs across the United States. The USPTO currently offers a number of tools and training events to assist and train small businesses about their foreign patenting opportunities and obligations. The SBA, too, has networks and resources in the small business community to reach the relevant audience. Moreover, the USPTO has previous experience partnering with SBA, as well as with other U.S. government agencies that work to support the needs of small businesses in internationalization, such as the Minority Business Development Agency and the International Trade Administration.

One notable program that the USPTO has conducted for over five years is the Intellectual Property Awareness Campaign (IPAC). This program provides day-long training sessions in various cities throughout the U.S focusing on intellectual property “basics” including patents, trade secrets, trademarks, domain names, and copyright protection and enforcement, both within the U.S. and abroad. The program is directed at small to medium-sized enterprises (SMEs), which often lack the training, knowledge, and resources possessed by larger companies. Because the USPTO offers the program on request in a web-based seminar format, this training can be tailored to the IP issues most critical to a particular type of small business. The web-based format also enables to the USPTO to reach many more small businesses in a manner, and at a time and place, which suits their needs. In 2011, the IPAC program had over one hundred attendees in various cities throughout the U.S. The IPAC education programs were delivered in person by teams of 4–6 USPTO staff per program.

The USPTO has considerable experience with developing and delivering educational outreach programs in partnership with other U.S. Government institutions. The USPTO’s current partners include the SBA; bureaus of the Department of Commerce including the Minority Business Development Agency and the International Trade Administration; and the U.S. Intellectual Property Enforcement Coordinator at the White House Office of Management and Budget. The USPTO has also worked with the U.S. Chamber of Commerce and the National Association of Manufacturers to educate small businesses about domestic and foreign IP protection through on the StopFakes.gov campaign, administered by the International Trade Administration. Not least, the USPTO has developed university partnerships, such as the IP Empowerment Summit at Howard University to educate minority-owned small businesses about domestic and international intellectual property law.

Ideally, any USPTO-SBA education and training program would be scalable so that it may reach more small businesses. One approach to expanding IP education and training for small businesses would be to increase the scale of the current IPAC program, using partnership arrangements and thereby increasing participation by small businesses. That could be accomplished by using electronic resources (such as internet based webinars) and by using physical locations such as the SBA-funded Small Business Development Centers (SBDC) (900 service locations) and the Patent and Trademark Resource Centers (79 locations) around the United States. Other non-government partners could be enlisted. The USPTO and SBA would be responsible for developing materials and training local businesses on day-long or modular small-business foreign patenting content. The USPTO and SBA could conduct more webinars with the existing infrastructure, and leverage the training materials already developed for the IPAC, as well as other patent-relevant courses, to reach larger audiences of small businesses.

Alternatively, the USPTO and SBA could partner, through the SBDCs, to expand upon the services that the SBDC's 900 locations currently provide to small businesses. While the SBDC currently provides training and counseling services to small business throughout the U.S., the USPTO and SBA could "train-the-trainer" so that local staff could deliver relevant materials on international IP protection, and its relation to internationalization strategies of small businesses, to a more diverse and larger segment of U.S. small businesses. These training sessions could be either in-person training or web-based distance learning. The goal would be to educate and train a significantly larger number of U.S. small businesses per year than the IPAC is currently able to reach.

3.The USPTO and SBA Should Engage Industry to Discuss How Best to Support U.S. Small Business Efforts to Patent Internationally

Several comments recognized that small businesses are a part of a larger economic ecosystem, and often engage, partner, and form alliances with different-sized companies at different stages of the innovation cycle and during their internationalization process. Often small companies enter into business relationships, including research and development alliances or distribution agreements across industries. Larger companies, too, often serve as investors in young startup companies through corporate venture capital. The health of the small business sector can be an important determinant of industry success, especially in those industries like pharmaceuticals and electronics that use acquisition of small businesses to fuel large-company innovation cycles.

Because many members of the public expressed a preference for market solutions over direct government intervention, it would be useful and informative for the USPTO, the SBA, and other allied agencies to convene discussions among industry representatives and small-business investors on small business foreign patenting.

4. The USPTO and SBA Should Collect More Information and Conduct More Study on the Most Appropriate Methods of Supporting International Patenting by Small Businesses

Public commentary – from a small sample of eighteen sources – voiced uncertainty about the advisability of the U.S. government becoming involved in subsidizing small business foreign patenting in a manner that would alter the operation of efficient market solutions. Moreover, there is virtually no evidence, either from academic research or from respondents, about the relative advantages of a loan versus a grant scheme as each relates to defraying the expenses of small businesses seeking international patent protection. Because too little meaningful data could be found relating to the advisability of the U.S. government using taxpayer funds either as a loan or a grant program to support this activity, USPTO does not recommend any program of public financing of small business foreign patenting at this time. However, it would be useful and informative as a next step for the USPTO, the SBA, and other allied agencies to collect more data.

As a general matter, the United States is stronger when we make investments in building the analytical capacity necessary to study the innovation economy. Intangible assets such as patents have been shown in one recent economic study to comprise over 50 percent of all U.S. business outputs,[88] and our capacity as a nation and government to analyze and make sense of these new sources of American national competitive advantage is vital to our long term economic health. The relative paucity of useful information and meaningful analysis concerning key questions that Congress asked the USPTO and SBA to study indicates that more investment in basic understanding of important questions related to American competitiveness in the innovation economy may be desirable. Meaningful policy evaluation requires both the best data and the best analytical capabilities focused on the important questions concerning how America will win the future.

Given the lack of information regarding the questions posed by Congress in the legislation, follow-up surveys and economic studies may be conducted to provide the legislature with the best information upon which to base its policy choices. Surveys targeted to America's small businesses, both those that currently use the patent system as well as innovating small companies that are opting out of the patent system, can provide useful information about their understanding of the value of international patenting and the utility of different means of providing support. Moreover, economic studies focusing on the relative advantages of choosing grant programs or loan programs, or other approaches to supporting international patenting for those small companies that need that need this protection may be helpful.

Pilot programs too can be an effective means of gathering information when issues are complex, and when there is considerable uncertainty about the benefits that may flow from the use of different possible policies. The USPTO has determined that further study is needed on these issues. In the near term the USPTO does not recommend that a new pilot program of support for international patenting be instituted, thereby avoiding additional burden on taxpayers until more information is gathered.

CONCLUSION

America's economic health is now and will continue to be determined by the health of America's small businesses. Because many small businesses rely on patent protection to thrive and grow, it is important to understand the relationship between success in the global economy and foreign patenting protection. Foreign patent protection can preserve valuable options for innovative U.S. small businesses to internationalize, to grow, and to become the job creators of tomorrow. The set of proposed recommendations contained in this report is a first responsive step toward better serving U.S. small businesses while gaining a better understanding of the needs and incentives of America's small business entrepreneurs with respect to international patent protection.

End Notes

[1] See Magnus Henrekson & Dan Johansson, Gazelles as Job Creators - A Survey and Interpretation of the Evidence (Research Institute of Industrial Economics, Working Paper No. 733, 2008), available at http://ssrn.com/abstract=1092938.

[2] See Tim J. Kane, The Importance of Startups in Job Creation and Job Destruction (July 2010), available at http://ssrn.com/abstract=1646934.

[3] See E.J. Reedy & Robert E. Litan, Starting Smaller; Staying Smaller: America's Slow Leak in Job Creation (July 2011), available at http://ssrn.com/abstract=1883660.

[4] Leahy-Smith America Invents Act, Pub. L. No. 112-29, § 31, 125 Stat 284, 339-40 (2011).

[5] Request for Comments and Notice of Public Hearings on the Study of International Patent Protection for Small Businesses, 76 Fed. Reg. 62,389, 62,391 (Oct. 7, 2011).

[6] See Public Comments for International Protection for Small Business Study, available at http://www.uspto.gov/aia_implementation/intl_patent_protection.jsp, Transcript of Public Hearing on the Study of International Patent Protection for Small Businesses: Hearing Before USPTO & SBA (Oct. 27, 2011) [hereinafter "USPTO Hearing"], available at http://www.uspto.gov/ aia_implementation/111027-ipsb_transcript.pdf, and Transcript of Public Hearing on the Study of International Patent Protection for Small Businesses: Hearing Before USPTO & SBA (Nov. 1, 2011) [hereinafter "USC Hearing"], available at http://www.uspto.gov/ aia_implementation/ 111101-ipsb_transcript.pdf.

[7] Comments of Biotechnology Industry Organization, at 3, available at http://www.uspto.gov/aia_implementation/ipp-2011oct20-bio.pdf.

[8] Patent Technology Monitoring Team, USPTO, U.S. Patent Statistics Chart, Calendar Years 1963–2010, available at http://www.uspto.gov/ web/offices/ac/ido/oeip/taf/us_stat.htm.

[9] Analysis of USPTO data. This measure uses the nation of origin of the first named inventor on the application as an indicator of domestic and foreign origin.

[10] The Kauffman Firm Survey (KFS) is a data source derived from new businesses founded in 2004 and tracked each year, providing information on these businesses from startup to sustainability, with data across their early years focusing on subjects such as how businesses are financed; the products, services, and innovations these businesses possess and develop; and the characteristics of their owners and operators. Alicia Robb & E.J. Reedy, An Overview of the Kauffman Firm Survey: Results from 2009 Business Activities (Apr. 5, 2011), available at http://ssrn.com/ abstract=1802597.

[11] See generally David DesRoches et al, Kauffman Firm Survey Baseline Methodology Report (Oct. 24, 2007) available at http://ssrn.com/abstract=1024045 (final report to the Ewing Marion Kauffman Foundation).

[12] Kauffman reports that the technology categories are based on the businesses' Standard Industry Classification (SIC) code designation, developed in the early 1990s by researchers at the Bureau of Labor Statistics. See Paul Hadlock et al, High Technology Employment: Another View, 114 MONTHLY LAB. REV. 26 (1991). The figures are consistent with the findings from earlier years of the survey, when these companies were younger. See Kauffman Foundation, New Kauffman Foundation Study Offers Insights into the Earliest Years of a New Business (March 12, 2008), available at http://www.kauffman.org/ Details.aspx?id=1090.

[13] See Christian Helmers & Mark Rogers, Does Patenting Help High-tech Start-ups?, 40 RES. POL'Y 1016 (2011) (finding UK firms obtaining patents in the early 2000s to have significantly higher growth in total assets over the following five years), and Ronald J. Mann & Thomas W. Sager, Patents, Venture Capital, and Software Start-ups, 36 RES. POL'Y 193 (2007) (finding that U.S. venture-backed software firms holding patents in the late 1990s showed better later performance in venture capital financing based on total investment, how the firm exited, and how long the firm survived).

[14] See Stuart J.H. Graham et al, High Technology Entrepreneurs and the Patent System: Results of the 2008 Berkeley Patent Survey, 24 BERKELEY TECH. L.J. 255 (2009).

[15] Ronald J. Mann & Thomas W. Sager, Patents, Venture Capital, and Software Start-ups, 36 RES. POL'Y 193 (2007); David H. Hsu & Rosemarie H. Ziedonis, Strategic Factor Markets and the Financing of Technology Startups: When Do Patents Matter More As Signaling Devices? (June 2011), available at http://wwwmanagement.wharton.upenn.edu/hsu/ inc/doc/papers/david-hsu-signaling.pdf. It is noteworthy that an inherent limitation of this kind of evidence, whether statistical or anecdotal, is that it is difficult to distinguish which way the causation runs. For instance, the growth, superior performance, and success of a small company may have been primarily the result of good managers, who also tend to select patenting because it is in their best interest to preserve the company's competitive options.

[16] See, e.g., Joachim Wagner, The Causal Effects of Exports on Firm Size and Labor Productivity: First Evidence from a Matching Approach, 77 ECON. LETTERS 287 (2002), Jane W. Lul & Paul W. Beamish, The Internationalization and Performance of SMEs, 22 STRATEGIC MGMT. J. 565 (2001), and Antonio Majocchi & Antonella Zucchella, Internationalization and Performance: Findings from a Set of Italian SMEs, 21 INT'L SMALL BUS. J. 249 (2003).

[17] Exec. Order No. 13,534, 75 Fed. Reg. 12,433 (Mar. 11, 2010).

[18] Exec. Order No. 13,534, 75 Fed. Reg. 12,433 (Mar. 11, 2010).

[19] See, e.g., Zoltan J. Acs et al, The Internationalization of Small and Medium-Sized Enterprises: A Policy Perspective, 9 SMALL BUS. ECON. 7 (1997).

[20] See, e.g., Ariel Pakes, Patents as Options: Some Estimates of the Value of Holding European Patent Stocks, 54 ECONOMETRICA 755 (1986).

[21] Comments of American Intellectual Property Law Association, at 4, available at http://www.uspto.gov/aia_implementation/ipp-2011nov08-aipla.pdf.

[22] Comments of David Carlson, at 2, available at http://www.uspto.gov/aia_implementation/ipp-2011oct24-carlsondavid.pdf.

[23] See generally, StopFakes.gov, http://www.uspto.gov/smallbusiness/about/faq.html (last visited Jan. 13, 2012).

[24] USC Hearing, at 32, available at http://www.uspto.gov/aia_implementation/111101-ipsb_transcript.pdf (Q&A with Christopher Palermo).

[25] See, e.g., Zoltan J. Acs et al, The Internationalization of Small and Medium-Sized Enterprises: A Policy Perspective, 9 SMALL BUS. ECON. 7 (1997).

[26] Comments of Philip McGarrigle, at 4, available at http://www.uspto.gov/aia_implementation/ipp-2011nov01- mcgarrigle-philip.pdf.

[27] Comments of American Intellectual Property Law Association, at 3, available at http://www.uspto.gov/aia_implementation/ipp-2011nov08-aipla.pdf.

[28] Comments of American Intellectual Property Law Association, at 3, available at http://www.uspto.gov/aia_implementation/ipp-2011nov08-aipla.pdf.

[29] Comments of Power Clean 2000, Inc., at 1, available at http://www.uspto. gov/aia_implementation/ipp2011nov09-power-clean-2000.pdf.

[30] Comments of Power Clean 2000, Inc., at 1, available at http://www.uspto.gov/ aia_implementation/ipp2011nov09-power-clean-2000.pdf.

[31] Comments of David Carlson, at 1, available at http://www.uspto.gov/aia_implementation/ipp-2011oct24-carlsondavid.pdf.

[32] Testimony of Dr. Jay Kesan, at 2, available at http://www.uspto.gov/aia_implementation/ipp-2011oct31-kesanjay.pdf.

[33] Mogee Research & Analysis Associates, Foreign Patenting Behavior of Small and Large Firms, Final Report to the Small Business Administration under Contract No. SBA-8140 (1996), available at http://archive.sba.gov/advo/research/rs167.html.

[34] Mary Ellen Mogee, Foreign Patenting Behavior of Small and Large Firms: An Update, Report to the Small Business Administration under Contract No. SBAHQ-01-M-0357 (2003), available at http://archive.sba.gov/advo/research/rs228_tot.pdf.

[35] Comments of Christopher Palermo, at 1, available at http://www.uspto.gov/aia_implementation/ipp-2011nov01- palermo-christopher.pdf.

[36] Comments of American Intellectual Property Law Association, at 3, 5, available at http://www.uspto.gov/aia_implementation/ipp-2011nov08-aipla.pdf.

[37] Comments of Biotechnology Industry Association, at 3, available at http://www.uspto.gov/aia_implementation/ipp-2011oct20-bio.pdf.

[38] Comments of Philip McGarrigle, at 2, available at http://www.uspto.gov/ aia_implementation/ipp-2011nov01- mcgarrigle-philip.pdf.

[39] Comments of American Intellectual Property Law Association, at 4, available at http://www.uspto.gov/aia_implementation/ipp-2011nov08-aipla.pdf.

[40] Comments of David Carlson, at 1, available at http://www.uspto.gov/aia_implementation/ipp-2011oct24-carlsondavid.pdf.

[41] Comments of David Carlson, at 1, available at http://www.uspto.gov/aia_implementation/ipp-2011oct24-carlsondavid.pdf.

[42] See USPTO Hearing, at 21–24, available at http://www.uspto.gov/ aia_ implemen tation/111027- ipsb_transcript.pdf.

[43] See USPTO Hearing, at 24–26, available at http://www.uspto.gov/aia_implementation/ 111027- ipsb_transcript.pdf.

[44] As of 2011, the USPTO charges three preliminary fees to small entities: a $95 filing fee, $310 search fee, $125 examination fee (totaling $530). A small business is required to pay another fee of $875 fee at patent issue. The USPTO also requires that, in order to keep the patent in force, the patentee must pay post-grant maintenance fees at staggered intervals, equaling $565 at 31/2 years from issuance, $1,425 at 71/2 years from issuance, and $2,365 at 111/2 years from issuance. See USPTO Fee Schedule, http://www.uspto.gov/curr_fees (last visited Jan. 13, 2012). Moreover, the AIA mandates reduced fees for micro-entities. See Pub. L. No. 112-29, § 10(b).

[45] As a comparison point, the following calculations are based on a patent application in several countries containing 23 claims (allowing a comparison to the USPTO basic filing fee which allows 3 independent claims and 20 dependent claims). The following calculations compare fees paid for such a typical application at the USPTO with fees that would be paid by a U.S. small business at the European Patent Office (EPO), the Japan Patent Office (JPO), and the China State Intellectual Property Office (SIPO).

At the European Patent Office (EPO), for instance, preliminary fees for applicants include a €105 filing fee ($134 in 2010), a €1105 search fee that includes Europe ($1,510 in 2010), a €1785 international search fee ($2274 in 2010), and a €1480 examination fee ($1,885 in 2010), totaling a 2010 equivalent of $5803, more than ten times the equivalent cost for similar services for small businesses in the USPTO. See European (EPC) Fees, http://www.epo.org/applying/forms-fees/fees.html (last visited Jan. 13, 2012).

The JPO charges a ¥24,000 filing fee ($262 in 2010), and an examination fee of ¥118,000 ($1,292 in 2010) plus ¥4,000 per patent claim ($44 in 2010), which in the stylized example of 23 claims would equal a 2010 U.S. dollar equivalent of $2,561 for all these preliminary costs, about five times more expensive than these services offered to small entities at the USPTO. See JPO Schedule of Fees, http://www.jpo.go.jp/tetuzuki_e/ryoukin_e/ryokine.htm (last visited Jan. 13, 2012).

[46] As a comparison point, the following calculations are based on a patent application in several countries containing 23 claims (allowing a comparison to the USPTO basic filing fee which allows 3 independent claims and 20 dependent claims).

The EPO requires annual fees, so-called annuities, both during the application phase and after a patent is allowed. These fees amount to €420 a year beginning at year three, €525 at year four, €735 at year five, €945 at year six, €1,050 at year seven, €1,155 at year eight, €1,260 at year nine, and €1,420 at years ten and greater. See European (EPC) Fees, http://www.epo.org/applying/forms-fees/fees.html (last visited Jan. 13, 2012).

At the JPO, annual renewal fees are ¥2,300 plus ¥200 per claim for years 1–3, ¥7,100 plus ¥500 per claim for years 4–6, ¥21,400 plus ¥1,700 per claim for years 7–9, and ¥61,600 plus ¥4,800 per claim for year 10 and greater. See JPO Schedule of Fees, http://www.jpo.go.jp/tetuzuki_e/ryoukin_e/ryokine.htm (last visited Jan. 13, 2012).

Similarly, the Chinese SIPO requires annual maintenance fees of RMB1,200 per year for years 4–6, RMB2,000 per year for years 7–9, RMB4,000 per year for years 10–12, RMB6,000 per year for years 13–15, and RMB8,000 per year for year 16 and thereafter. See Schedule of Fees for Chinese Patent (All-China Patent Agents Association), http://www.liu-shen.com/docs/SFBEN.pdf (last visited Jan. 13, 2012).

[47] Dietmar Harhoff et al, Languages, Fees and the International Scope of Patenting (Centre for Econ. Pol'y Res., Discussion Paper No. 7241, 2009), available at http://www2.druid.dk/conferences/viewpaper.php?id=5704&cf=32.

[48] Comments of Christopher Palermo, at 1, available at http://www.uspto.gov/aia_implementation/ipp-2011nov01- palermo-christopher.pdf.

[49] USPTO Hearing, at 28 (testimony of Morgan Reed), available at http://www.uspto.gov/aia_implementation/111027-ipsb_transcript.pdf.

[50] Comments of Christopher Palermo, at 2, available at http://www.uspto.gov/aia_implementation/ipp-2011nov01- palermo-christopher.pdf.

[51] Comments of American Intellectual Property Law Association, at 2, available at http://www.uspto.gov/ aia_implementation/ipp-2011nov08-aipla.pdf.

[52] USPTO Hearing, at 52 (testimony of Timothy King), available at http://www.uspto.gov/aia_implementation/111027-ipsb_transcript.pdf.

[53] USPTO Hearing, at 52–53 (testimony of Timothy King), available at http://www.uspto.gov/aia_implementation/111027-ipsb_transcript.pdf.

[54] USPTO Hearing, at 53 (testimony of Timothy King), available at http://www.uspto.gov/aia_implementation/111027-ipsb_transcript.pdf.

[55] Comments of Christopher Palermo, at 1, available at http://www.uspto.gov/aia_implementation/ipp-2011nov01- palermo-christopher.pdf.

[56] See generally Paul A. Gompers & Joshua Lerner, THE VENTURE CAPITAL CYCLE (2004).

[57] Dane Stangler, High-Growth Firms and the Future of the American Economy, March 2010 (Kauffman Foundation)

[58] The U.S. government currently offers financial support through myriad programs to small businesses. For instance, the 2011 launch of the Startup America program included several initiatives including the SBA-led launch of two $1 billion initiatives for impact investing and early-stage seed financing. See Startup America: Obama Administration Commitments, http://www.whitehouse.gov/issues/startup-america-public (last visited Jan. 13, 2012).

[59] Comments of Christopher Palermo, at 1, available at http://www.uspto.gov/ aia_ implementation/ipp-2011nov01- palermo-christopher.pdf.

[60] See Stuart J.H. Graham et al, High Technology Entrepreneurs and the Patent System: Results of the 2008 Berkeley Patent Survey, 24 BERKELEY TECH. L.J. 255 (2009).

[61] See, e.g., AMERICAN INTELLECTUAL PROPERTY LAW ASSOCIATION, 2011 REPORT OF THE ECONOMIC SURVEY (2011).

[62] See, e.g., Anne M. Scheiderman, "Filing International Patent Applications under the Patent Cooperation Treaty (PCT): Strategies for Delaying Costs and Maximizing the Value of Your Intellectual Property Worldwide" in IPHANDBOOK OF BEST PRACTICES § 10.7, available at http://www.iphandbook.org/handbook/ch10/p07/.

[63] See generally Jean O. Lanjouw et al, How to Count Patents and Value Intellectual Property: The Uses of Patent Renewal and Application Data, 46 J. INDUS. ECON. 405 (1996).

[64] USPTO Hearing, at 65 (testimony of Alan Kasper), available at http://www.uspto.gov/aia_implementation/111027-ipsb_transcript.pdf.

[65] USPTO Hearing, at 98–99 (testimony of Steven Caltrider) , available at http://www.uspto.gov/aia_implementation/111027-ipsb_transcript.pdf.

[66] USPTO Hearing, at 35 (Q&A with Morgan Reed) , available at http://www.uspto.gov/aia_implementation/111027-ipsb_transcript.pdf.

[67] Act of Aug. 27, 1982, Pub. L. No. 97–247, § 1, 96 Stat. 317 (1982), implemented by 47 Fed. Reg. 40,134, 40, 139–40 (codified as amended, at 37 C.F.R. §§ 1.27, 1.28 (1983)).

68 Ulead Sys., Inc. v. Lex Computer & Mgmt. Corp., 351 F.3d 1139, 1142 (Fed. Cir. 2003).

[69] Act of Dec. 12, 1980, Pub. L. No. 96–517, § 2, 94 Stat. 3015, 3017–18 (1980).

[70] H. Rept. 112–98 (report of the House Committee on the Judiciary), available at http://www.gpo.gov/fdsys/pkg/CRPT-112hrpt98/pdf/CRPT-112hrpt98-pt1.pdf.

[71] A "micro-entity" is defined in pertinent part as an entity or person who has not been named as an inventor or more than 4 previously filed U.S. non-provisional patent applications and whose income does not exceed 3 times the U.S. median household income. Leahy-Smith America Invents Act, Pub. L. No. 112–29, §§ 10(b), 10(g) (2011).

[72] Patent Rules of Canada, SOR/96-423 § 3.01(3), available at http://laws-lois.justice.gc.ca/eng/regulations/SOR-96- 423/index.html. Furthermore, small entities cannot be controlled by any entity that employs more than 50 employees and is not a university, and cannot be under an obligation of assignment or license in the invention to such an entity.

[73] See http://www2.aipla.org/html/patent-handbook/countries/mexico/MXgeneral.html.

[74] Wu Zhongze, Vice Minister of Science and Technology, People's Republic of China, Innovation: China's New National Strategy, Address at the Opening Ceremony of China-EU Science and Technology Year (Oct. 11, 2006), available at http://ec.europa.eu/research/iscp/eu-china/pdf/vm_wu_speech_en.pdf.

[75] State Intellectual Property Office, People's Republic of China, National Patent Development Strategy (2011–2020), available at http://graphics8.nytimes.com/packages/pdf/business/SIPONatPatentDevStrategy.pdf.

[76] Wen Jainchun et al, Research on Patent Fees Subsided by Local Government in China, in PROC. INT'L CONF. ON INFO. MGMT., INNOVATION MGMT. & INDUS. ENGINEERING (2008), available at http://ieeexplore.ieee.org/stamp/stamp.jsp? tp=&arnumber=4737773.

[77] Zhen Lei et al, Patent Subsidy and Patent Filing in China, Address at the Conference on Innovation and Patent Harmonization (Sept. 30–Oct. 1, 2011).

[78] Jason Cooper & Stephanie Chu, SURGE IN CHINESE INNOVATION AND THE IP IMPLICATIONS (2009), available at http://www.ipo.org/AM/Template.cfm? Section =Patents&Template=/CM/ContentDisplay.cfm&ContentID=28247.

[79] Comments of American Intellectual Property Law Association, at 6, available at http://www.uspto.gov/aia_implementation/ipp-2011nov08-aipla.pdf.

[80] Comments of Power Clean 2000, at 2, available at http://www.uspto.gov/aia_implementation/ipp-2011nov09- power-clean-2000.pdf.

[81] Comments of Jay Kesan, at 3–4, available at http://www.uspto.gov/aia_implementation/ipp-2011oct31-kesanjay.pdf.

[82] Comments of Power Clean 2000, at 2, available at http://www.uspto.gov/ aia_implementation/ipp-2011nov09- power-clean-2000.pdf.

[83] Comments of American Intellectual Property Law Association, at 7–8, available at http: //www.uspto.gov/aia_implementation/ipp-2011nov08-aipla.pdf.

[84] Comments of Paul Overhauser, at 2, available at http://www.uspto.gov/aia_implementation/ipp-2011oct27- overhauser-paul.pdf.

[85] Comments of Christopher Palermo, at 2–3, available at http://www.uspto.gov/aia_implementation/ipp2011nov01-palermo-christopher.pdf.

[86] Comments of Biotechnology Industry Association, at 3, available at http://www.uspto.gov/aia_implementation/ipp-2011oct20-bio.pdf.

[87] Comments of American Intellectual Property Law Association, at 5, available at http://www.uspto.gov/aia_implementation/ipp-2011nov08-aipla.pdf.

[88] See Carol Corrado et al, Intangible Capital and U.S. Economic Growth, 55 REV. INCOME &WEALTH 661 (2009).

In: Patent
Editor: Shawn Bennett
ISBN: 978-1-63483-626-5

Chapter 2

REPORT ON THE PRIOR USER RIGHTS DEFENSE*

U.S. Patent and Trademark Office

EXECUTIVE SUMMARY

The Leahy-Smith America Invents Act (AIA), signed into law on September 16, 2011, represents the most substantial overhaul of U.S. patent law in generations. Among the fundamental changes the AIA makes to the U.S. patent system—and a central feature of the new law—is the transition from a first-to-invent to a first-to-file system of awarding patents.

The AIA also expands the "prior user rights" defense to infringement and broadens the classes of patents that are eligible for the new limited prior user rights defense. Appropriate delimited prior user rights protect third parties who can demonstrate that they were commercially using the an invention for at least one year prior to the filing date of a patent application relative to such invention. U.S. law already provided a prior user rights defense that was limited to patents directed to methods of conducting business. The AIA, by contrast, extends the prior user rights defense to patents covering all technologies, not just business methods. At the same time, the AIA includes several limitations and exceptions to the prior user rights defense, including a

* This is an edited, reformatted and augmented version of a report to Congress issued January 2012.

prohibition against license, assignment or transfer of the defense, other than in connection with an assignment or transfer of the entire business to which the defense relates; the defense is geographically limited to cover only those sites where the invention was used before the critical date; and there is an explicit exception to the defense for patents owned by or assigned to universities or affiliated technology transfer organizations. These limitations and exceptions to the prior user rights defense address the equitable interests of patentees, universities and affiliated organizations.

In response to concerns expressed during the legislative debate about the impact of an expanded prior user rights defense on the patent system, and innovation more generally, Congress mandated the United States Patent and Trademark Office (USPTO) to produce a report detailing its findings and recommendations on the operation of prior user rights in the industrialized world. Specifically, Congress directed the USPTO to examine and report on the following issues:[1]

A. A comparison between patent laws of the United States and the laws of other industrialized countries, including members of the European Union and Japan, Canada, and Australia.
B. An analysis of the effect of prior user rights on innovation rates in the selected countries.
C. An analysis of the correlation, if any, between prior user rights and start-up enterprises and the ability to attract venture capital to start new companies.
D. An analysis of the effect of prior user rights, if any, on small businesses, universities, and individual inventors.
E. An analysis of legal and constitutional issues, if any, that arise from placing trade secret law in patent law.
F. An analysis of whether the change to a first-to-file patent system creates a particular need for prior user rights.

Pursuant to this mandate, the USPTO held a public hearing and solicited comments from interested parties to complement its own independent research on the issue of prior user rights. Additionally, the USPTO obtained input from the Office of the United States Trade Representative (USTR), the Department of Justice (DOJ), and the Department of State (DOS). The USPTO received 29 comments from a broad spectrum of foreign and domestic stakeholders, including industry organizations, universities, bar associations, and individuals. Most of the comments supported the AIA's prior user rights defense, though several expressed concerns about the effect such a defense could have on, *inter alia*, patent valuation and the prompt disclosure of new innovations. In this respect, a number of comments discussed the interaction between patents and trade secrets as a business strategy for protecting inventions. While a few comments took a view that prior user rights may have the unintended consequence of promoting secrecy over disclosure in a manner that could be detrimental to the patent system, many others noted that the ability to maintain trade secrets is vital to American competitiveness and job growth, and that a limited prior user right defense is an appropriate complement to a first-to-file system.

On the basis of these comments, our own research, and input provided by USTR, DOJ, and DOS, the USPTO finds that:

1) The AIA strikes the right balance by limiting the prior user rights defense to those parties that can prove commercial use at least one year prior to the filing date of the patent application by clear and convincing evidence.
2) The scope of the prior user rights defense includes limitations on the type of continued activities, the transfer of the personal rights, and the enforcement of said right, such that the patentee's rights are not unjustly impinged and the university community may benefit, since the defense is not available in patent actions by universities.
3) There is not substantial evidence that prior user rights as established in the AIA will have a negative impact on innovation.
4) There is no substantial evidence showing that prior user rights as established in the AIA will have a disproportionately negative impact on venture-capital investments to small businesses and start-ups.
5) There is an insufficient basis to recommend a change to the scheme chosen by Congress with respect to the application of prior user rights to universities.
6) There is no substantial evidence that the limited prior user right defense as established by the AIA will have a negative impact on small businesses or independent inventors.
7) A prior use defense to patent infringement is both Constitutional and lawful and the defense is consistent with the Constitution and Supreme Court precedent recognizing that trade secret law and patent law can, and do, legally co-exist in the United States, and indeed have co-existed since our Constitution was created.

8) Trade secret protection is of considerable value to United States businesses and the United States economy, and as such, there are compelling economic and policy justifications for providing a prior user rights defense to patent infringement.
9) Providing limited prior user rights in a first-inventor-to-file system addresses the inherent inequity such a system creates between an earlier commercial user of the subject matter and a later patentee. A prior user rights defense is pro-manufacturing and pro-jobs, as it rewards businesses that put new technology promptly into commercial use, and provides protection for early commercial use when challenged by the later filing of patent applications by other entities.
10) Because the availability of a prior user rights defense to patent infringement is a fundamental aspect of many patent regimes throughout the industrialized world, there is a strong preference that United States businesses be afforded the same advantages in terms of prior use protections in the United States that their competitors enjoy abroad.

Accordingly, pursuant to the congressional mandate under AIA[2], the USPTO recommends the following:

1) The prior user rights defense provisions set forth in the AIA are generally consistent with those of major trading partners and need not be altered at this time.
2) The prior user right defense under the AIA should be maintained with no change at the present time because there is no substantial evidence that it will have a negative impact on innovation, venture funding, small businesses, universities, or independent inventors.
3) The USPTO should reevaluate the economic impacts of prior user rights as part of its 2015 report to Congress on the implementation of the AIA, when better evidence as to these impacts might be available.
4) United States patent law should provide for a prior user rights defense as an appropriate balance between trade secret protection and patent law, which legally co-exist to provide competitive advantages for United States businesses.
5) United States patent law should provide for a prior user rights defense to patent infringement in order to address a systemic inequity inherent in a first-inventor-to-file system and to ensure United States businesses are (i) able to protect their investments in the event of a

later issued patent, and (ii) placed on similar footing as competitors in other jurisdictions.

I. INTRODUCTION

On September 16, 2011, President Obama signed into law the AIA,[3] a landmark piece of legislation representing the most comprehensive overhaul of United States patent law in generations.[4] The AIA ensures that the United States patent system fulfills the Constitutional imperative to "promote the Progress of Science and useful Arts"[5] in light of 21st century economic and business challenges, while also maintaining American leadership in the global innovation marketplace.[6] To this end, the AIA streamlines our patent system and improves the quality of patents granted in the United States to minimize unnecessary litigation costs.[7]

For example, the AIA enhances protection available for non-patented innovations and trade secret assets. This is best illustrated in the expansion of the defense to patent infringement based on earlier invention and use, commonly referred to as "prior user rights" or, more appropriately, the "prior user rights defense."[8] A "prior user rights defense" generally refers to a limited defense to patent infringement afforded to a party that was commercially using, or engaged in substantial preparations for commercial use of, an invention that was later patented by another party.[9] This defense acknowledges the fact that inventors may, for a variety of reasons, prefer not to seek patent protection.[10] By allowing the earlier user/inventor to continue uninterrupted commercial use of the invention while also allowing the later inventor to obtain a patent enforceable against all others, the prior user right embodies a balance between the equitable economic interests of the earlier user against the larger goal of the patent system.[11] This larger goal of the patent system is the enrichment of the fund of human knowledge.[12] In so doing, the prior user right also may further the aim of the AIA in reducing litigation and related costs.[13]

The prior user rights defense to patent infringement is a common fixture in patent systems of many countries around the world,[14] and is not without precedent in the United States.[15] Prior to the enactment of the AIA, for instance, section 273 of Title 35 provided for a limited prior use defense applicable to business method patents. Section 273 of Title 35 is available in Appendix A. Section 273 was enacted as part of the 1999 American Inventors Protection Act (AIPA)[16] in response to the decision by the Court of Appeals for the Federal Circuit (CAFC) in *State Street Bank & Trust Co. v. Signature*

Financial Group, Inc.,[17] which affirmed that business methods are eligible subject matter for patenting.[18] In doing so, however, the decision itself created uncertainty for United States businesses as to whether they might now be liable for patent infringement for continued use of internal business processes once thought to be unpatentable.[19] The *State Street* decision also followed several years of debate in Congress and academic and legal circles about introducing a broad prior user rights defense into United States patent law.[20] The language of section 273 reflected this confluence of circumstances by, on the one hand, providing an explicit provision for a prior user rights defense, and on the other hand, narrowly circumscribing its applicability to patents covering methods of doing or conducting business.

The AIA makes a number of significant changes to the existing prior user rights defense. Most notably perhaps, the AIA expands the defense to all technologies, not just to business methods.[21] More specifically, the defense is available to persons who independently commercially employed the invention in the United States in connection with an internal commercial use, an arm's length sale, or an arm's length transfer of a useful end result of the commercial use.[22] To be entitled to the defense, however, the prior user must establish that the relevant activities occurred more than one year before the earlier of (1) the filing date of the patent application; or (2) the date of public disclosure by the patentee during the patentee's grace period.[23]

Importantly, the AIA provides for several limitations and exceptions to the new prior user rights defense. For example, the defense is personal, and thus may not be licensed, assigned, or transferred, other than in connection with an assignment or transfer of the entire business to which the defense relates.[24] In addition, the applicability of the defense is geographically limited to cover only those sites where the invention was used before the critical date.[25] The AIA further provides an explicit exception to the defense for patents owned by or assigned to universities or affiliated technology transfer organizations.[26] This latter provision relates to concerns expressed by the higher education community that prior user rights may impair the ability of universities to license patents on upstream research results.[27] Appendix B features the prior user rights provision of the AIA.

The AIA will serve to protect trade secret owners and other alleged infringers with a defense against patent infringement. This means that, in applicable cases, keeping your invention as a trade secret, such as a process that is used to make a product that is not disclosed to the world, is less risky for the business community. If someone invents a manufacturing process that improves or optimizes the efficiency, cost, or quality of producing a particular product, the inventor has the choice of either keeping the details of the process as a trade secret or file for patent protection. Without a prior use defense, the risk of going down the trade secret path is that a competitor who independently arrives at the same invention may obtain a patent and sue for infringement even though the prior user invented the enhancements and had been using it commercially for a long time.

As with the creation of the prior user rights defense in the AIPA, the limited expansion of the defense in the AIA also sparked debate among the patent and business communities. Proponents argue that the defense is necessary in a competitive economy and strikes the balance between trade secret and patent protection. Proponents likewise argue a prior user rights defense is necessary in a first-inventor-to-file regime to provide inventors who put inventions to practice first with the ability to continue using their innovations, even though they may not be entitled to a patent. Critics argue, however, that prior user rights undermine the purpose of the patent system by creating a strong incentive to protect innovations as trade secrets rather than disclose them and fuel technological growth in the United States.

Given the opposing views on the prior user rights defense, Congress mandated in section 3 of the AIA that the USPTO produce a report on the operation on prior user rights in industrialized countries in order to explore the implications of a prior use defense in the United States. In particular, Congress required the USPTO to study the following topics, in consultation with the Office of the U.S. Trade Representative (USTR), Department of State (DOS), and Department of Justice (DOJ):

A. A comparison between patent laws of the United States and the laws of other industrialized countries, including members of the European Union and Japan, Canada, and Australia.
B. An analysis of the effect of prior user rights on innovation rates in the selected countries.
C. An analysis of the correlation, if any, between prior user rights and start-up enterprises and the ability to attract venture capital to start new companies.
D. An analysis of the effect of prior user rights, if any, on small businesses, universities, and individual inventors.
E. An analysis of legal and constitutional issues, if any, that arise from placing trade secret law in patent law.
F. An analysis of whether the change to a first-to-file patent system creates a particular need for prior user rights.[28]

This Report addresses each of the topics identified by Congress and analyzes "prior user rights" in the context of changes made by the AIA. This Report first provides background in Section II on the scope of the study that the agency undertook to fulfill its Congressional mandate. Next, the Report provides an international comparison of prior user rights in Section III, Part A, and then explores the impact of prior user rights on various innovation activities, including their relationship to start-ups, universities, and independent inventors in Section III, Part B. The Report examines the relationship between trade secret protection and patent law in Section III, Part C, as well as the interaction between the prior use defense and the provisions of the AIA that create a first-to-file patent system in the United States in Section III, Part D. When appropriate, the Report also provides findings and recommendations based on the analysis set forth in each section.

II. BACKGROUND

To fully appreciate the role of prior user rights in the international arena and better understand the impact of a prior user rights defense on innovation and businesses, the USPTO gathered information and perspectives through a solicitation of public comments via a Federal Register Notice posted October 7, 2011.[29] The Federal Register Notice was posted on the USPTO Web site and the distributed to various intellectual property law organizations for dissemination to their members. Interested members of the public were invited to give oral testimony at a hearing scheduled for October 25, 2011, and/or submit written comments on issues related to the operation of prior user rights. The Notice specifically requested the public to share their experiences and opinions with respect to prior user rights in foreign jurisdictions.[30] Mirroring the language of the AIA, the Notice also asked the public to provide information with respect to the effect of prior user rights on innovation rates, start-up enterprises, small businesses, universities, and individual inventors.[31] Additionally, the Notice requested any comments on the legal analyses pertaining to the placing trade secret law in patent law and the interaction between the prior use defense and adoption of a first-inventor-to-file system in the United States.[32]

On October 25, 2011, the USPTO held a public hearing on these issues at the USPTO in Alexandria, Virginia. Five witnesses provided testimony, including Alan Kasper, Immediate Past President of the American Intellectual Property Law Association; Gary Griswold, Consultant and Chair Emeritus of the Coalition for 21st Century Patent Reform; Thomas Kowalski, Shareholder, Vedder Price, P.S.; Dan Lang, Vice President Intellectual Property and Deputy General Counsel, CISCO Systems, Inc.; and MaCharri Vorndran-Jones, Intellectual Property Law Section, American Bar Association. Subsequently, the USPTO received a total of 29 written comments from large and small companies, independent inventors, leading patent practitioners, the university community, academics, patent law scholars, and professional organizations.[33] Appendix C provides a listing of entities that provided input to the USPTO through oral testimony and/or written comments.

To complement the public submissions and in order to ensure an appropriate foundation of information on which to base the analyses contained in this Report, the USPTO also conducted an extensive literature review and consulted with patent officials from foreign patent offices. In addition, as required by Congress, the USPTO consulted with the USTR, the DOS, and the DOJ in compiling this Report.[34]

This Report, along with the findings and recommendations contained herein, are based on public comments, literature reviews, and independent legal research. All of the information gathered for this Report afforded the USPTO the opportunity to fairly evaluate the operation and impact of a prior use defense to patent infringement.

III. DISCUSSION

A. Comparison of Prior Use Provisions in Industrialized Countries

The first inquiry of this congressionally mandated study is a comparison between the patent laws of the United States and those of other industrialized countries ("international comparison") regarding the availability and scope of prior user rights. The mandate expressly identifies the following industrialized countries: the European Union, Japan, Canada, and Australia. In addition to the identified countries, this Report also takes into consideration other industrialized (and emerging) countries and key trading partners of the United States: the Republic of Korea, Brazil, Mexico, the People's Republic of China ("China"), and the Russian Federation.

At the outset, the European Union does not, *per se*, encompass a common set of patent laws.[35] Instead, the patent system in Europe is governed by the European Patent Convention (EPC), which is a treaty outside the umbrella of treaties forming the European Union. The EPC does not establish a prior user rights regime. Thus, each member state of the European Union has its own unique prior user rights laws.[36] Accordingly, this Report focuses on a sample set of countries from the European Union that reflects different dimensions of populations, trade, and domestic economies. This sample set from the European Union includes: Denmark, France, Germany, and the United Kingdom.

Eight of twenty-nine public comments contained a discussion of particular experiences with prior user right regimes in foreign jurisdictions. Four of the eight were from entities based in other countries, including two from Australia,[37] one from New Zealand,[38] and one from Germany.[39] Additionally, one comment was from an international organization of patent practitioners.[40] Each of these submissions expressed the belief that prior user rights are important to achieving an equitable balance of interests, particularly from the perspective of a first-to-file system, though the degree of actual experience detailed in these submissions in terms of asserting or objecting to prior user rights varied substantially. Two of the comments, for instance, noted that those entities and/or their members had very little or no practical experience.[41] The other three comments, however, indicated that although there may be little litigation on prior user rights in certain jurisdictions, clients are counseled with some regularity on their applicability in particular situations.[42]

The remaining three of the eight comments were submitted by U.S.-based entities—two by companies representing different technology sectors[43] and the other from an organization representing a range of U.S. patent owners and stakeholders.[44] Both of the companies were supportive of prior user rights as a general matter, though, as with several of the above-mentioned comments from international stakeholders, neither indicated a significant degree of practical experience with exercising them.[45] The comments of the organizations were generally critical of prior user rights and, while acknowledging that none of its members had ever utilized prior user rights, noted that many of the arguments advanced by proponents are not supported by empirical evidence or other data.[46]

Eight other submissions touched on international aspects of prior user rights, but did not directly identify any particular experiences in applying them. These submissions were from a range of stakeholders—individuals, companies, and coalitions—and generally focused on the need to adopt prior user rights in the United States in order to level the international playing field for U.S. innovators and promote job growth in America.[47] A few of the comments, in particular, noted that the absence of prior user rights in the

United States could give an advantage to foreign competitors that may have cheaper labor costs or are subsidized by foreign governments.[48]

On balance, the public comments suggest that prior user rights are not problematic in other jurisdictions, and in fact, are a well-accepted practice. Only a few of the submissions detailed any actual experiences with prior user rights in foreign countries, and each of these was generally positive; none of the comments identified any first-hand, negative experiences. Moreover, several of the comments specifically noted that there has been little or no litigation of prior user rights in other countries, which suggests they are not creating any systemic issues.

Separate from the public's anecdotal evidence on the use of prior user rights in foreign jurisdictions, the USPTO conducted an analysis of the prior user rights laws of the above-mentioned countries, focusing on the necessary requirements to assert the prior user rights defense and the scope of said prior user rights defense. A summary of the USPTO's findings about prior user rights law in foreign countries is provided in Appendix D.

Analysis and Findings

1. Requirements to Assert the Prior User Rights Defense

For the most part, the requirements that a party must satisfy to qualify for a prior user rights defense include: (a) qualifying prior use activities; (b) qualifying prior user; (c) qualifying time period; and (d) qualifying locations. The following is a summary and comparison of the differing national requirements.

a. Qualifying Prior Use Activities

A primary inquiry for qualifying for the benefit of prior user rights laws is determining the type of activity that can legally be deemed as a prior use activity. With respect to the countries in this study, prior use ranges from prior commercial use, to mere possession of an invention.[49] In many instances, a hybrid form of the two is incorporated (e.g., use and substantial preparations for the same).[50] In addition, even within a single country, prior use may be defined differently based on the type of invention at issue. That is to say, an activity constituting prior use as to a product invention can sometimes fail to qualify as a prior use activity as to a process invention.[51] Of course, this requirement varies from country-tocountry. Still, despite the differences, the overarching narrative regarding prior use is that the legally qualifying activity is dependent on the actions of a prior user and/or the type of invention at issue.

With regard to the level of activity by a party asserting prior user rights, prior use activities can be categorized as either: (i) pure use; (ii) pure possession; or (iii) use-possession hybrids. The jurisdictions with a pure "use" definition are Brazil and Canada. The only jurisdiction with a pure "possession" definition is France. The rest of the studied countries adopt a use-possession hybrid form of prior use. Even between these use-possession hybrid regimes, a main point of divergence is the threshold for determining "possession."

Brazil's patent laws limit prior use activity to pure use.[52] More specifically, prior use activity in Brazil is defined as "exploitation" of the patented subject matter.[53] Accordingly, non-use of the invention, or even marginal use (e.g., preparations for the eventual use thereof), may act as a *de facto* waiver of a party's right to assert a prior user rights defense during litigation. Similarly, prior use in Canada is defined as "the purchase, construction, or acquisition" of a later patented invention.[54]

In contrast, prior use in France is limited to "possession of the invention."[55] The particulars as to what constitutes "possession of the invention" are not fully settled. [56] Generally speaking, however, preparatory studies are insufficient to establish prior user rights, but preparations of an industrial/productive nature would suffice to show "possession of an invention."[57]

The rest of the studied countries have a use-possession hybrid definition of prior use. For example, in Denmark, activities that constitute prior use include: (i) exploiting the invention commercially; and (ii) making substantial preparations for commercial exploitation of the same.[58]

Likewise, in Australia, prior use is defined as either: (i) exploiting the product, method, or process; or (ii) taking definite steps to do the same. Similar definitions are set out in the laws of Japan,[59] the Republic of Korea,[60] Mexico,[61] China,[62] and the Russian Federation.[63]

Some of these use-possession hybrid countries afford varying possession standards. For example, the United Kingdom has a very stringent possession standard. The United Kingdom defines prior use as "acts that would constitute infringement" of a patent or "effective and serious preparation" to do the same.[64] With regard to making "effective and serious preparation," the United Kingdom's case law suggests that mere possession of an invention, without proactive steps to manufacture or commercialize the same, is not enough activity to be deemed prior use.[65] In contrast, Germany has an easily satisfied possession standard. In Germany, activities that constitute prior use include "use of the invention" and making "necessary arrangements to do the same." [66]

With regard to the latter, a party asserting the defense must show actual possession of the invention.[67] Possession of the invention simply means that the asserting party must be knowledgeable of the invention and be able to intentionally exploit the invention.[68] This low threshold is circumscribed by the fact that prior use in Germany is limited to business or enterprise activities of a defendant.[69] Thus, prior use in Germany must be independent of any third parties and solely in the interest of the party asserting a prior user rights defense.[70] The use of a third party to realize the invention is not considered a prior use activity.[71] However, production for others is permissible.[72]

In view of the above, the prior user rights defense of the AIA appears to adopt a use definition similar to most of the studied countries. Under the AIA, prior use activity is defined as commercial use, "either in connection with an internal commercial use or an actual arm's length sale or other arm's length commercial transfer of a useful end result of such commercial use."[73] The statute also affirmatively sets forth several scenarios in which subject matter is explicitly considered as being "commercially used" and thus eligible for the prior user rights defense.[74] For example, one provision states that subject matter for which commercial marketing or use is subject to a premarketing regulatory review period during which safety or efficacy is established shall be deemed to be commercially used for purposes of the prior user right defense during the regulatory review period.[75] Additionally, a second provision provides that use of the subject matter by a "nonprofit research laboratory or other nonprofit entity, such as a university or hospital, for which the public is the intended beneficiary" will be considered a "commercial use" for purposes of the prior user right defense.[76]

It is noteworthy to briefly discuss in this study how in some jurisdictions, prior use may have different definitions based on statutory subject matter. These prior use definitions can be divided into three types: (i) activities related to product-inventions; (ii) activities related to product-by-process inventions; and (iii) activities related to process inventions. Of these, activities (i) and (ii) are usually coextensive.

For example, in Australia, the prior use standard is exploitation (and taking definitive step towards exploitation) of "the product, method or process in the patent area."[77] Exploitation, per the Australian standard, is defined differently based on the subject matter of the invention.[78]

Product-inventions and product-by-process inventions encompass a significant amount of activities under the rubric of *exploitation.*[79] One of the many activities under this nomenclature is offering to sell. In contrast, exploitation with respect to a pure process invention is limited to actual use only.[80] Thus, offering to sell a process invention does not trigger prior user rights. Similar to Australia, the countries of Japan[81] and the Republic of Korea[82] have different meanings based on the subject matter of an invention.[83] It should be noted, however, that most of the other studied countries do not have this piecemeal approach, and apply the same "use" standard across the board.

Unlike Australia, prior use activity under the AIA is not defined differently based on statutory subject matter. That is to say, the same "commercial use" standard is applied to all statutory subject matter, specifically defined in the AIA as, "subject matter consisting of a process, or consisting of a machine, manufacture, or composition of matter."[84]

b. Qualifying Prior User

A second inquiry for qualifying for the benefit of prior user right laws is "who" is a prior user. This inquiry focuses on ascertaining whether a prior user has to be the actual inventor, or alternatively, if a prior user may encompass someone who acquired the invention from another. In most countries, anyone who acquired an infringing invention, prior to the date of filing of the patent application, may assert a prior user rights defense. Most of these countries also have a requirement of good faith, discussed more fully below.

A requirement of invention in a prior user rights regime means that only an earlier inventor may assert prior user rights. Under such a system, prior user rights originally vest in a prior inventor. From that point, the right (or defense) can be transferred according to the legal standard of a particular country (discussed further below). The only country with a pure invention requirement appears to be Russia.[85]

According to Russia's patent law, any person who, prior to the date of filing, had "conceived and was using in good faith" an invention may seek benefit of prior user rights.[86]

In contrast, those jurisdictions that permit acquisition in their prior user rights regime allow anyone who has acquired the invention, through a purchase or conveyance, may assert prior user rights. It is noteworthy to indicate that, with the exception of Russia, every country studied permits prior user rights-based acquisition.[87] For example, in the United Kingdom, a party claiming benefit of prior use rights need only *perform any act* that would otherwise constitute infringement.[88] Thus, the prior user right is not contingent on prior invention of the subject matter.[89]

In view of the above, it appears that the AIA also has a standard consistent with the large majority of the studied countries in permitting acquisition of prior user rights. More specifically, the AIA states that the prior user rights defense "may be asserted only by the person who performed or directed the performance of the commercial use [...], or by an entity that controls, is controlled by, or is in common control with such a person."[90] In other words, the defense can be asserted by anyone that would be directly, or indirectly, liable of infringement to a patent holder of the same invention.[91] The AIA does not require that a prior user be an actual inventor.

Further, the presence of a good faith requirement helps determine whether a party may be disqualified from asserting prior user rights. All of the countries subject to this study have a good faith requirement in one form or another (i.e., statutorily, administratively, or jurisprudentially), with the exception of Mexico. Mexican law is silent on this issue of good faith, and the study did not find any administrative or judicial mention of good faith with respect to prior user rights. Good faith varies in definition, such that some countries' laws articulate specific acts that defeat the prior user right, while others have a general statement of "good faith" requirement. The latter, thus, may encompass inequitable activities that extend beyond copying.

With respect to the narrowly tailored good faith requirements, Germany, Denmark, France, Australia, and Japan provide insight as to the variance between these articulated good faith definitions. For example, German patent law includes a provision where if a party enters into a non-disclosure agreement with an eventual patent holder, the party is expressly required to wait six months before asserting a prior use defense against the same patent holder, assuming the patent holder did not previously file a patent application.[92] Thus, how a party obtained knowledge of an invention (either from a patentee, third party, or independently) is of little value in Germany.[93] In fact, merely learning of a patent applicant's invention is not, *ipso facto*, sufficient grounds to support an allegation of bad faith.[94] A party would have to also show breach of contract and/or misappropriation, among other things.[95] Another example of a narrowly tailored good faith requirement is Denmark, where prior user rights will only exist "provided that the exploitation did not constitute *an evident abuse in relation to* the [patent] applicant or his legal predecessor."[96] In Australia, a person may not claim benefit of prior user rights if he/she "derive[d] their invention from the patent holder, or patentee's predecessor," unless that information was voluntarily made available by the patent holder.[97] Finally, in Japan good faith is defined as "*a lack of knowledge of the content of a patent application*."[98] Thus, this sample set of good faith standards illustrate how varied good faith standards can be from one jurisdiction to another.[99]

In contrast, some countries have a general good faith requirement. These general requirements do not articulate a specific activity that would bar a party from asserting prior user rights. As a consequence, this general good faith requirement is much broader. For example, France's prior user rights law states that any person who satisfies the various elements for claiming benefit of prior user rights, and acted "in good faith," may assert such a right.[100] Specific activities that would constitute (or destroy) good faith are not articulated. Nonetheless, practitioners understand that this broad standard encompasses various inequitable activities, such as acquiring an invention in breach of a contractual obligation.[101] Other countries that have a generalized good faith requirement are the United Kingdom,[102] Brazil,[103] and the Russian Federation.[104]

Some countries do not specifically articulate a good faith requirement in their laws, but incorporate similar standards through some other vehicle. Canada, for example, does not articulate any good faith requirement. However, Canadian courts have interpreted its patent law to include some form of "good faith" requirement. [105] While Canadian courts may not prevent acquisition of prior user rights based merely on what might be considered "dishonorable" behavior, they have disallowed prior user rights when the prior use activity approaches fraud. As such, the Canadian prior user rights regime requires some level of "good faith," although it is based on case law rather than on a statutory requirement.

Likewise, China's prior user rights laws do not articulate a specific good faith requirement.[106] Nonetheless, there is at least a minimal good faith requirement in China imposed by the courts. [107] In particular, an alleged infringer cannot claim the prior user right defense on the basis of illegally obtained technology or designs.[108] Mexico is also another country that does not specifically articulate a good faith requirement in the law, and it is unclear whether Mexico imposes a good faith requirement.[109]

The AIA has both a general good faith requirement, and also an articulation of specific conduct that would defeat the prior user rights defense. The general articulation of the good faith requirement is that a prior user must be, "acting in good faith, [to] commercially [use] the subject matter in the United States."[110] In addition to this general requirement, the AIA goes on to further articulate a specific activity. More specifically, that "[a] person may not assert a defense under this section if the subject matter on which the defense is based was derived from the patentee or persons in privity with the patentee."[111] Thus, generally speaking, the AIA is consistent with the countries examined in this Report on two fronts, by including a generalized and an articulated good faith standard.

c.Qualifying Time Period

A third inquiry in qualifying for the benefit of prior user right laws is "when" must prior use activity take place. By definition, the prior use activity must take place *before* a specific date related to the grant of a patent. Thus, this inquiry focuses on said specific date. In this regard, the temporal requirement of all the studied countries is generally consistent, with some exceptions as noted below. For all the studied countries, the prior use activity must take place before the earliest effective filing date of the patent application.[112] For example, in the United Kingdom, the activity constituting prior use must take place before the priority date of an invention.[113] According to the *U.K. Patent Act*, the priority date of an invention is defined as the date that an application was filed, or the date that the earliest parent application was filed.[114]

Germany's prior user rights laws, however, have special provisions that directly affect this otherwise consistent temporal requirement. In Germany, prior use activity must take place at, or before, the date of filing.[115] Like the rest of the studied countries, a patentee can claim priority to a parent application.[116] However, this rule does not apply to nationals of a foreign country that does not have reciprocity laws.[117] Therefore, foreigners from said countries are limited to the actual filing date of an application. That is to say, a prior user need only beat a national filing date, as opposed to priority date related to a foreign application from a jurisdiction without reciprocal laws.

With respect to the AIA, the U.S. further builds upon the temporal requirement of the studied countries. In order to assert the prior user rights defense under the AIA, commercial use must have occurred at least one year before the earlier of either: "(A) the effective filing date of the claimed invention; or (B) the date on which the claimed invention was disclosed to the public in a manner that qualified for the exemption from the prior art under section 102(b)."[118] This makes the U.S. temporal approach significantly more restrictive than that for any other prior user rights system.

d. Qualifying Locations

The final inquiry in qualifying for the benefit of prior user rights laws is "where" prior use activity must take place. More specifically, prior user rights laws articulate a specific geographical or political area where prior use activities will be legally recognized as triggering the benefit of the same set of laws. Like the previous inquiry, most studied countries have consistent provisions regarding the locality of prior use.[119] More specifically, the prior use must take place within the borders of each respective country. Thus, activities taking place within a foreign country are not taken into account in determining whether the prior use defense is available.

For example, in Japan, any activity constituting prior use must take place in Japan.[120] This territorial requirement may also span territories owned by a particular country. For example, like Japan, France requires prior use to take place within its borders as well.[121] Since France also encompasses territories outside continental Europe, this area includes overseas departments such as French Polynesia, the Wallis and Futuna Islands, the French territories of the Southern Hemisphere, Antarctic, New Caledonia, and Mayotte.[122]

Denmark, France, Germany, and the United Kingdom are all members of the European Union. Therefore, it should be noted that some scholars have indicated that limiting the territorial scope of prior user rights to within the national borders of member states may contravene European Competition Law.[123]

The AIA is in harmony with all studied countries, as it provides that only commercial use in the United States may provide the basis for a prior use defense.[124]

Thus, the USPTO finds that:

Finding 1: The AIA strikes the right balance by limiting the prior user rights defense to those parties that can prove commercial use at least one year prior to the filing date of the patent application by clear and convincing evidence.

2. Scope of Prior User Rights Defense

The scope of the prior user rights defense varies somewhat from country to country. The following is a summary and comparison of the scope of the defense, which in general is fairly consistent across the vast majority of the jurisdictions studied. The scope of the prior use defense can generally be characterized as: (a) breadth of the prior use activities encompassed by the defense; (b) transferability of the prior use defense; and (c) limitations on enforceability of the defense.

a. Breadth of the Prior Use Activities Encompassed by the Defense

The breadth of the prior user activities among all the studied countries is fairly consistent. Generally, the prior user right laws of these countries protect or allow for the continuation of activity that was commenced before the effective filing of patent. While this "right" may be somewhat nuanced in its characterization (as either a defense, a non-exclusive license, or an exemption from liability), the practical effect is mostly uniform throughout the jurisdictions studied. The breadth of protection afforded to a prior user can be classified into one of the following three categories: (i) commensurate scope; (ii) broader scope; or (iii) narrower scope.

A little more than half of the studied countries require that the prior user limit any continued activity to a scope commensurate with the previous activity that triggered the prior user rights. For example, in Denmark, those claiming prior user rights may continue exploiting the invention as long as said exploitation "retain[s] its general character."[125] In Brazil, prior users may "continue the exploitation, without onus, in the same manner and under the same conditions as before."[126] Similarly, prior user rights in the Russian Federation provide "a right to proceed with [...] prior use free of charge, provided that the scope of the activity [...] is not extended." [127] The same appears to be true for Australia,[128] Canada,[129] Mexico,[130] and China.[131]

A second group of the studied countries provide a broader scope of prior use activities. To be more accurate, the "broadening" is actually an extension of the original prior use activities to include any equivalents or variants of the original activity. For example, the Republic of Korea provides for prior user rights in the form of a "non-exclusive license." The breadth of this non-exclusive license is limited to "the scope of the objective of the invention or the business related to the invention that the person is working or making preparations to work."[132] The prior user may expand the business based on the prior user right, and may change the mode of operation to the extent that those skilled in the art are expected to employ the changed mode.[133] Japan also has a similar provision.[134]

France and the United Kingdom also provide a broader scope for prior use activities, but do so in a different fashion. The prior user rights laws of the United Kingdom do not explicitly broaden the scope of activities afforded to a prior user, but only allow the prior user to continue with the "infringing acts."[135] However, the United Kingdom's case law has interpreted this to mean that the prior user may expand a business activity based on his/her prior user right and may change the mode of operation if such a change does not affect the essence of the invention.[136]

Similarly, the French prior user rights statute does not place any explicit limits on the scope of the prior user right.[137] However, French Courts have interpreted this legislation to broaden the scope of the personal right, such that it includes subject matter identical to the patented invention, as well as any equivalents of the patented invention.[138] It should be noted, however, that unlike the United Kingdom's case law precedent, France's case law is non-binding, as it is based on a civil law tradition.

The final category regarding the breadth of prior use activities include laws that actually narrow the scope of permissible prior use activities when compared to the triggering prior use. Germany is the only studied country with such laws. As previously discussed, Germany has a hybrid use-possession prior use system, in which a party must show either use of the invention or "necessary arrangements" to use the invention. This triggering activity is not circumscribed any further. However, once the benefit of prior user rights is afforded, a prior user must limit any continued use of the invention to "the needs of his own business in his own workshops or the workshops of others."[139] That is to say, the scope of prior user rights is narrowed to exclude non-commercial activity and even commercial activity that is not related to the specific needs of a business. In Germany, a party that intends to continue infringing use, based on a claim of prior user right, bears the burden of showing a nexus between said continued infringing activity and some particular business need.

The breadth of permissible prior use activities under the AIA is of a broadening nature, which is similar in about half of the studied countries. The AIA specifies that the prior user rights defense, "extends only to the specific subject matter for which it has been established that a commercial use [...] occurred."[140] However, the provision goes on to state that "the defense shall also extend to variations in the quantity or volume of use of the claimed subject matter, and to improvements in the claimed subject matter that do not infringe additional specifically claimed subject matter of the patent."[141] In permitting the prior user rights defense to be invoked even where there are variations in the quantity or volume of the use, the AIA permits a greater breadth of prior use activities than in some other jurisdictions.

b. Transferability of the Prior Use Defense

In most jurisdictions, the prior use defense is limited to prior users and may only be transferred or assigned along with some or all of a business entity utilizing the prior user right. Prior user rights generally cannot be licensed, and as such, are typically treated as a personal right. The limits on transfer and prohibition of licensing keeps the prior use defense an equitable defense rather than an affirmative right. Among the studied countries, transferability of the prior user rights defense can be classified as: (i) transferable; (ii) transferable only with a business; (iii) transferable with either a business or a production unit; or (iv) not transferable.

Only one of the studied countries appears to have an unqualified transferability requirement. Prior user rights in Australia can be transferred to "successors in title."[142] More specifically, Australian law permits assignment of prior user rights. However, prior user rights cannot be licensed by the prior user. Assignees are considered as "successors in title" while licensees are not considered as such because assignees are subsequent owners of rights.[143] Australian law does not articulate that there needs to be a nexus between assignment and a business unit of the original holder of the prior user rights.

Most of the studied countries permit the transfer of prior user rights if done in conjunction with the transfer of the business originally holding such a right. For example, Denmark limits the transfer of prior user rights by requiring that the right "shall only be transferred to others together with the business in which it has arisen or in which the exploitation was intended."[144] Germany[145] and China[146] have similar laws in this respect.

Japan also limits transfer of prior user rights to an assignment in conjunction with a business. However, Japan also provides other avenues of transfer. Generally speaking, Japanese prior user rights may be transferred whenever the business involving "the working of" the relevant invention is also transferred. But in addition to this, the prior user right may be transferred where the consent of the patent holder is obtained or where the transfer occurs as a result of general succession (including inheritance).[147] The same is true for the Republic of Korea.[148]

The third category is transferability of the right alongside a business or a production unit of said business. More specifically, rather than limiting conveyance of said prior user right to an assignment in conjunction with a whole business, said conveyance may include an assignment in conjunction with the particular business unit that is responsible for the triggering prior use activity. This type of transfer is actually of a lower threshold than the preceding category. For example, French prior user rights can "be transferred together with the business, the enterprise or the part of the enterprise to which it belongs."[149] Transfer of prior user rights is also similarly restricted in the Russian Federation, where the right of a prior user "may be transferred together with the production unit in which actual use or necessary preparations for said use have been made." [150] Brazil[151] has laws similar to France and the Russian Federation.

The United Kingdom also has similar but perhaps even more restrictive provisions on transfer of prior user rights as compared to France, the Russian

Federation and Brazil. Prior users may assign or transmit their right upon death to any person who acquires that part of the business.[152] Furthermore, the United Kingdom allows a prior user in a partnership to authorize his partners to perform the prior use activities.[153]

The final category bars the transfer of the prior user right. While Mexican law does not appear to explicitly address transfer of prior user rights, it is understood in practice that prior user rights are not transferable.[154] As such, prior user rights appear to be a purely personal and non-transferable right in Mexico.[155]

Finally, it should be noted that Canada was not classified into any of the categories above. As previously stated, Canada's prior user right encompasses the purchase, construction, or acquisition of claimed subject matter.[156] Therefore, any party in a product supply chain, or business transaction, may also be able to assert the prior user rights defense. However, it is not clear whether the right may be transferred to (and asserted by) someone outside of this nexus. Canadian law appears silent on this issue and Canadian case law does not appear to provide a definitive conclusion, with some suggesting that it is logical that they can be "passed on to another."[157]

Like most of the studied countries, the prior user rights defense under the AIA may be transferred as long as it is in conjunction with a business unit. More specifically, the AIA specifies that prior user rights can only be transferred, except to the actual patent owner, as part of a good faith assignment or transfer of the entire enterprise or line of business to which the defense relates.[158] That is, the right to assert said defense may not "be licensed or assigned or transferred to another person except as an ancillary and subordinate part of a good-faith assignment or transfer for other reasons of the entire enterprise or line of business to which the defense relates."[159]

c. Limitations on Enforceability of the Defense

Under section 273(e)(5) of the AIA, a prior user right defense to infringement may not be asserted if "the claimed invention with respect to which the defense is asserted was, at the time the invention was made, owned or subject to an obligation of assignment to either an institution of higher education... or technology transfer organization whose primary purpose is to facilitate the commercialization of technologies developed by one or more such institutions of higher education."[160]

After studying the prior user rights provisions on the various selected countries across the globe, there appears to be no limitations on enforceability in any of their laws. In this regard, it appears that the United States is rather unique is including a provision benefiting the academic sector.

Thus, the USPTO finds that:

Finding 2: The scope of the prior user rights defense includes limitations on the type of continued activities, the transfer of the personal rights, and the enforcement of said right, such that the patentee's rights are not unjustly impinged and the university community may benefit, since the defense is not available in patent actions by universities.

On the basis of the above analysis and findings, the USPTO recommends as follows:

Recommendation 1: The prior user rights defense provisions set forth in the AIA are generally consistent with those of major trading partners and need not be altered at this time.

B. Impact of Prior User Rights on Innovation, Start-up Enterprises, Small Businesses, Individual Inventors and Universities

The American Invents Act mandates that the USPTO conduct a prior user rights study, requiring the Office to conduct several inquiries related to economic activity.[161] This section addresses:

A. The effect of prior user rights on innovation rates in the selected countries;
B. If there is any correlation between prior user rights and start-up enterprises and the ability to attract venture capital to start new companies; and
C. The effect of prior user rights, if any, on small businesses, universities, and individual inventors.

Economic thought on the relationship of property rights and the incentives to engage in creative efforts like invention is well developed. Traditionally, economists have viewed patent rights as a solution to a basic failure in the market for ideas.[162] It has been recognized that ideas and creativity – and the inventions that the human mind may produce – have a character like that of a public good. Unlike a private good, a public good suffers from an excludability problem – meaning that it is difficult to exclude others from appropriating the economic benefits from the creator. As a result, without intervention, inventions would tend to be produced at a level lower than what is socially optimal since – without some kind of legal protection – some if not all of the economic value from creativity would flow to copiers, and not to the creator. As a result, creators would have little incentive to invest substantial amounts of time and capital in inventing because there would be little prospect of recouping research and development expenses. Patent rights work to solve this problem by providing a legal excludability for a limited time to the inventor, and by so doing work to provide the appropriate economic incentive to inventors to make the investments necessary for society to enjoy new technologies.

That said, the exclusivity provided by society to inventors is not complete. The Constitution enshrines the basic economics underlying the kind of right that society ought to offer to the inventor – that the exclusive right be "limited" and that it be tied to other ends associated with the "progress of science and the useful arts." Prior user rights tend to work against the excludability function of patents, but it is not clear from existing research how much and to what extent that may be a problem, even apart from the fact that existing research does not address the precise form of prior user rights provided under the AIA.

From an economic perspective, the availability of a prior user rights defense may offer both advantages and disadvantages. In terms of private benefits and costs, a preexisting inventor with a secret use of a later patented invention would be benefited by not being prevented from practicing the invention. Conversely, a later patenting inventor would be disadvantaged by having, at the margin, fewer opportunities to profit from the patented invention.

The benefits and costs to society, however, are more speculative, and will include the impact that the availability of the defense has at least upon innovation rates, the types of innovations that are produced, the speed and quality of that innovation, and the costs to consumers of the innovations available in the marketplace. These various costs and benefits are very unclear, and there is only limited economic research dealing with any of these implications.

Any meaningful disadvantages associated with the availability of the defense would be driven by two related effects: prior user rights could discourage actual innovation in society due to the lower economic returns available to inventors, and the defense could discourage the choice of patenting in relation to trade secret protection. Accordingly, it is relevant to ask whether any economic research has tried to determine how large these two effects are in the aggregate (in other words, what is the impact on innovation, disclosure, and follow-on innovation).

The remainder of this section includes a discussion of the three specific economic questions posed by Congress in light of these considerations, the public comments received, and the existing scholarly research and economic analysis available.

Analysis and Findings

1. Prior User Rights and Innovation Rates

This section considers the economic impact of prior user rights defense as it relates to innovation rates. First, it reviews the extent of prior user rights available under the AIA. Second, it discusses economic research related to the patent system and prior user rights, and their likely impacts on innovation. It concludes by analyzing the existing empirical research on the impact of prior user rights and presents findings.

As a threshold matter, this section discusses the economic impact of a prior user rights defense on innovation rates in selected countries (the United States, members of the European Union, Japan, Canada, and Australia).[163] As observed in comments to the Federal Register Notice from two associations of intellectual property attorneys, country-specific information is limited, with only occasional economic studies that focus on particular jurisdictions.[164]

Historically, prior user rights were included in the U.S. patent law between 1836 and 1952, and as noted in comments submitted to the Request For Information (RFI), no evidence has been generated that shows prior user rights weakening innovation rates or economic growth during that time.[165] In fact, industries were born and grew to prominence during that era, both those like petroleum in which secrecy tends to be a relatively more effective means of protecting innovation, and those like pharmaceuticals in which patenting is relatively more effective at allowing innovators to profit from the creation of new technologies.[166] A narrower prior user right was established by the American Inventors Protection Act of 1999 in section 273.[167] Since that time, it appears that only one case has been reported in which this defense has been raised.[168]

The Economics of Patent Rights and Innovation. There is very limited economic research on the incentive effects of a prior user rights defense and its impact on innovation. One economist used theoretical models to investigate these impacts, concluding that "[w]hen nearly simultaneous, independent invention occurs, awarding one inventor a patent and the other the right to use the invention has very attractive properties. Competition is enhanced, innovation is rewarded with relatively little deadweight loss, and the private and social incentives to be the sole versus joint inventor are generally better aligned than in the absence of such rights."[169] Another theoretical treatment concludes that independent invention defenses reduce entry into patent races, and thus reduce wasteful duplication, thereby preserving limited societal resources for other investments.[170]

Public comment to the Federal Register Notice was not consistent concerning the innovation impacts of allowing a prior use defense. Some members of the public warned of reduced innovation rates as a result of the availability of the defense: An organization representing patent owners said that prior user rights could harm innovation rates by reducing the confidence innovators have in the system of patent rights.[171] The same organization speculated that prior user rights may reduce the incentive to disclose innovations, by providing a form of protection for firms that maintain their innovations as trade secrets.[172]

In addition to newness, other characteristics of technology may play a role in the benefit and costs associated with a prior user defense. In some technologies characterized by numerous incremental inventions, it may not be feasible to patent every invention that a company is developing and using.

Comments made to the Federal Register Notice from two information technology companies said that patent protection is not the most efficient tool in some high-technology fields,[173] and that the patent system was not intended to force such firms to defensively patent everything.[174] Similarly, a pharmaceutical delivery company wrote that prior user rights were useful to defend against patents that provided only minor improvements and were used primarily to block generic competitors.[175]

The extent of benefits and harm flowing from the availability in the U.S. of a prior user rights defense is, in the end, an empirical question best answered by comprehensive data on the true impacts revealed over time. Unfortunately, formal statistical study of prior user rights has been extremely limited, largely because the full effect of this type of policy change is inherently difficult to measure. As a policy matter, the introduction of prior user rights often comes along with several other policy changes (as is the present experience of the United States in the AIA). A formal effort to measure whether the introduction of prior user rights affected innovation would thus be confounded by other policies changing simultaneously, which almost certainly also have an impact on innovation rates. Similarly, cross-country comparisons cannot effectively identify the effect of prior user rights, because prior user rights regimes often differ between countries, and countries always differ in many other ways, particularly in how their innovation systems are organized and the ways in which they operate.[176]

There is, however, some qualitative evidence available. There have been prior user rights in the United States for business method patents since the enactment of a 1999 law, but by all accounts this particular defense does not appear to have had a significant impact, either in industrial practice or in patent litigation. Also, the prior user rights available historically in the U.S. did not appear to have a significant impact on innovation.[177] Moreover, many countries have prior user rights in some form,[178] but their use in practice has been very limited.[179] Thus, after considering the available evidence, the USPTO finds that:

Finding 3: There is no substantial evidence that prior user rights as established in the AIA will have a negative impact on innovation.

2. Prior User Rights and New Company Venture Financing

In order to report possible correlations between the exercise of a prior user rights defense and the ability of new firms to attract venture capital, it is necessary to determine the drivers of venture-capital investment in the entrepreneurial economy. Anecdotal evidence, including comments submitted by several respondents to the RFI, suggests that patenting is often a critical element in venture investors' decision of whether to fund a technology start-up. As a tangible right covering an intangible asset, a patent can convey key information to potential investors. Patents may thereby provide both (i) information that investors may use to project potential commercialization success and returns on investment, and (ii) exclusivity rights that investors may use in estimating future market size.

Economic scholarship supports the patent-investment relationship. One particular economic study examined the relationship between venture funding and patenting rates across twenty industries and three decades[180], finding that venture funding was seven times more likely to stimulate patenting than was corporate R&D. The impact of venture capital on patenting rates for seed and early stage start-ups was especially significant, increasing the relative impact of a venture-funded dollar by 45% when compared to broader set of venture-funded firms. Overall, the study revealed that venture-backed firms were more likely than those without venture funding to engage in patenting, to have their patents cited, and to find themselves in "frequent and protracted litigation of both patents and trade secrets." Additionally, a growing body of research over the last ten years has indicated that patents are both important as selection criteria for venture capitalists[181] and are related to increased valuations for start-ups.[182] While many of the studies have focused on high-technology venture capitalists (VCs), who may value patent protection more strongly than in other sectors, this emphasis still aligns with the anecdotal evidence relating patent importance in valuation.

The core questions related to whether the availability of prior user rights negatively affect a start-up's ability to attract venture capital include: determining if potential use of these prior user rights is likely to either increase risk to or uncertainty for the venture capital investor; and to decrease the projected returns on investment.

While economic literature squarely addressing these questions is lacking, several public comments to the Federal Register Notice addressed the import of prior user rights to VC investment. Several respondents, including a prior director of a technology-focused Federal research organization,[183] asserted that prior user rights would create incentives for businesses to protect innovations through trade secrets instead of seeking patents. The implication is that the potential for secret uses in a market would make it more difficult for investors to estimate markets and value invention,[184] thereby depressing the attractiveness of a later patenting start-up to venture-capital funding. However, opinion was mixed on this question: One response by a major technology association which represents large, medium and small companies[185] suggested that, without the prior user rights defense, the company which lost the "race to the USPTO" in a first-to-file environment would lose all investment returns for that project. By avoiding such a harsh result in the particular case, the prior use right could provide venture capitalists with increased investment certainty.

A search of the existing economic literature yielded no empirical study of the impact of prior user rights within the U.S. venture-capital investment context, so correlations *per se* are not available. However, relevant information can be gleaned from a) identifying assertion rates and impacts of prior user rights in other countries, and b) assessing the degree to which trade secrets are already included in normal decision making by venture capital firms when making their start-up investments.

According to a recent European Commission study on prior user rights,[186] "very few cases involving prior use rights are brought to court [in Europe]," and questions posed to their stakeholders suggested that "the impact of prior use rights on patent litigation seem to be low." Europe shares with the U.S. many of the economic characteristics that ought to be relevant in assessing the impact of a prior user rights defense in the U.S.: It is highly innovative, supports a strong patent system, and possesses well-developed capital markets and a large and vibrant competitive environment. It is noteworthy, however, that Europe tends to have a less well-developed entrepreneurship, start-up, and venture-investing environment than does the U.S.

While Europe's experience with prior user rights may suggest that the U.S., too, will experience few impacts from the adoption of the defense, these data are not complete. Without the benefit of empirical evidence, it will not be possible to make determinations before U.S. companies have had a chance to exercise these rights, and legal precedent is developed as to when a prior use defense may validly be supported. Clarification of the extent of subject matter upon which a prior user rights defense can be exercised, specifically with regard to other commercial processes may ultimately help VC investors better estimate the risk of prior users actually affecting the return on their investments.

The best available evidence shows, however, that the U.S. venture-capital investment environment has grown and thrived in a context of uncertainty over property rights in innovation, especially as regards patents and trade secrets. In a survey conducted at Carnegie-Mellon University in the early 1990s by Wesley Cohen and coauthors, these researchers found that among U.S. manufacturing firms, the majority of industries studied considered secrecy as an effective means of appropriating returns from innovations.[187] The importance of secrecy to protecting innovation by U.S. companies since the early 1990s suggests that venture-capital firms have long experience estimating the risk in a competitive environment that contains a large number of trade secrets. Academic research conducted in 2008 reported on one experienced VC who said that:

> [w]hen thinking about the life-cycle of a company, in many respects the value of the IP is really generally assessed at the early stage by the first stages of "professional" money. These early-stage [professional investors] will do a great deal of scrutiny of IP of all types, but especially of trade secrets and patents.[188]

Such research suggests that investors have experience navigating the complexities of patenting and secrecy used by innovators in making investment decisions. Moreover, if prior user rights are likely to be rarely exercised in the U.S. (as is suggested by the European experience), then it is reasonable to conclude that "switching" from patent to trade secret protection as a result of the availability of this defense may be limited in nature.

Thus, after considering the available evidence, the USPTO finds that:

Finding 4: There is no substantial evidence that prior user rights as established in the AIA will have a disproportionately negative impact on venture-capital investments to small businesses and start-ups.

3. Prior User Rights and Small Businesses, Universities, and Individual Inventors

Due to inclusion of the America Invents Act university exception,[189] whereby the prior user rights defense cannot be asserted for inventions that were "made, owned, or subject to an obligation of assignment to either an institution of higher education ... or a technology transfer organization" which was primarily linked to higher education institutions, the potential impact of the defense on universities is more limited. Accordingly, this Report considers the impact of prior user rights upon universities separately from the impact on small businesses and inventors in the foregoing analysis.

a. Impact on Universities

Since the Bayh-Dole Act of 1980 changed the way in which universities and federally supported research laboratories interact with the patent system, a substantial body of economic research has developed investigating this relationship.[190] While this literature examines the impacts on university-sponsored invention and innovation in the wake of the increased opportunities for patenting, licensing, and commercialization of federally funded research ushered in by the Bayh-Dole Act, this research does not specifically deal with the relationship of U.S.-university patenting to the availability of a prior user right defense.

As mentioned previously, while prior user rights exist in all EU member states (except Cyprus),[191] reports show that there is a very low incidence of associated litigation. Furthermore, review of the scholarly literature did not reveal any study explicitly assessing the impact of prior user rights to institutions of higher education in Europe. It is not surprising that, given the low incidence of prior user rights cases generally, empirical studies on the impacts on a subset of innovative activity – the universities – from these countries is not readily available. Furthermore, public comments on the topic generally reflected a lack of data and anecdotal accounts of any negative impacts upon universities. For instance, a comment to the Federal Register

Notice from one large multinational company explicitly indicated that the company was unaware of any "evidence that universities in [European] countries had been negatively impacted by prior user rights."[192] Comments to the Federal Register Notice from a German patent attorney went further, noting a lack of negative effects, explaining that "small businesses, universities, and individual inventors in Germany are frequently SMEs, remote from the patent system," and that prior user rights protected their "legitimate know-how [from being] taken away."[193] Because many U.S. universities have a more direct involvement with the patent system, as often do their licensee start-up companies, this example from Germany may have less relevance to the U.S. case.

Commentary to the Federal Register Notice offered by several higher education associations[194] as well as a university-linked patent management organization,[195] expressed concerns about possible negative effects that a future strengthening of prior user rights could have upon innovation at universities in the US. These respondents recognized that prior user rights may be important in particular technology sectors, particularly those in which "complex products and manufacturing processes" may make it difficult to patent every component. However, they also expressed concern that extending prior user rights into the university arena could create disincentives for information disclosure and collaboration in the university environment (for instance, through peer-reviewed journals and conference presentations). Since innovations from universities tend to be basic research, early-stage, and highly risky, these responding associations expressed a concern that allowing prior use rights in the university environment could "expand the pool of trade secrets immune from assertion of patent rights," thereby reducing the certainty over what a patent excludes and discouraging private companies from licensing university patents.[196]

Congress mandated that all university and university-associated technology transfer organizations' patents will be exempted from the availability of a prior user rights defense (excepting non-Federally funded activities).[197] Accordingly, the associations' joint comments assert that the Act "effectively balances the concerns and interests of universities and private sector companies ... while inhibiting the development of a massive pool of trade secrets immune from the assertion of patent rights, as well as ... products that could be derived from early disclosures."[198] In summation, because prior user rights (and the university exclusion) have only recently been permitted to be widely available here in the U.S., there is insufficient evidence at this time to determine the impact on universities.[199]

Thus, after considering the available evidence, the USPTO finds:

Finding 5: There is an insufficient basis to recommend a change to the scheme chosen by Congress with respect to the application of prior user rights to universities.

b.Impact on Small Business and Individual Inventors

It is noteworthy that, as observed elsewhere in this study, evidence suggests that where prior user rights are available around the world, they are rarely used. These rights are rarely litigated in Europe,[200] in other countries where available around the world,[201] or under the business method version of prior user rights in the United States.[202] Indeed, in the case of business method patents in the United States, the prior user defense appears never to have been successfully raised since the enactment of the defense.[203] Since business methods are particularly prone to prior and independent invention,[204] such evidence of non-assertion may be particularly persuasive.

While scholarly research on the relationship of prior user rights defenses to small-business and independent-inventor incentives and performance is very limited, a few articles have been written. The scholarship is contradictory, however; some legal scholars raise a concern that small businesses may be harmed by the availability of a prior user rights defense, but other commentary suggests the contrary. One scholar has noted that it may be impossible to tell whether a possible infringer has prior user rights without expensive litigation that small businesses and independent inventors cannot afford.[205] Another suggested that small businesses would be unable to take advantage of the defense; small entities may have little or no manufacturing capabilities, and adopt licensing-only approaches, and thus are relatively unable to assert prior user rights.[206]

On the other hand, other scholarship has noted potential benefits to small businesses flowing from the defense. For example, some entities may license their rights to manufacturing or marketing firms, which themselves may qualify for prior user rights. If the licensing entity fails to obtain a patent—for example, because after the license it is discovered another entity filed first—its licensees may be able to continue to use the licensed technology, through prior user rights.[207]

In addition, not all small entities will adopt patent-oriented innovation strategies. Some companies invest in production facilities immediately, and given limited funding, may not be able to both pursue a growth strategy and patent their innovations simultaneously. Prior user rights may be a useful defense to these small entities. Indeed, to the extent that prior user rights disproportionately affect small entities, they may benefit due to their greater difficulty in patenting, especially in light of the fact that small firms might not have the resources to pay for litigation to enforce their patents.

Public comments to the Federal Register Notice on the impact of prior user rights legislation upon small businesses and inventors were diverse and did not present a coherent answer to the questions presented by Congress. Some comments supported the position that prior use rights do not have a harmful effect on small businesses and independent inventors, and others suggested that harsh economic effects may flow from the availability of the defense.

Joint comments submitted by six higher education associations[208] suggested that complex manufacturing and products may require hundreds (or thousands) of potentially patentable components, and that prior user rights in the commercial world could help innovators defend against a patent acquisition company. Such a company may pursue a strategy of not manufacturing, but instead attempting to extract profits from companies. In this way, a manufacturer could be forced to stop all production if it chose to leave a small component of an otherwise complex product unpatented.[209] A small business manufacturer could be affected by such an outcome. More commonly, however, small businesses will be component suppliers to larger manufacturers, and such an outcome would stop production at the larger company, thus affecting the small component supplier as well. Two major technology companies, each of which purchase components from many small businesses, stated that it simply was not feasible to patent every component and process involved in highly complex technology sectors, and that these large producers should not be penalized if another entity patented a portion of the processes they were previously using.[210]

Furthermore, a major communications association asserted that any lack of prior user rights in a first-to-file system, would "compel firms to aggressively seek patents on minor improvements despite the very limited public benefit."[211] In such a defensive "race to patent," even small innovations could increase patent filings with negative impacts on all innovators who wish to use a timely patent system, while decreasing the quality of the patents requested.

A group of green technology companies stated that they often maintain key innovations as trade secrets in order to defend their competitiveness, because it can be very difficult to detect if others use their patented processes (especially in an international manufacturing environment). [212] In this situation, they suggested that prior user rights would not create incentives to increasingly rely on trade secrets; instead it would decrease the risk of an external patent affecting the processes already in use. Similar sentiments came from a space transportation company, which relies on trade secrets in order to protect innovations from being copied outside of the U.S.[213] Additionally, both testimony[214] and submitted comments[215] from individuals in the IP law arena suggested that the prior user rights being available outside, but not inside, the U.S. could lead to the exporting of manufacturing. The degree to which such exporting affects large versus small companies was not explored.

Other comments collected during the Federal Register Notice and public hearing process suggested that small businesses and independent inventors would be harmed by the availability of the defense in the U.S. Comments from an independent inventor stated that most start-ups and inventors will not meet the America Invents Act requirement to have used their innovation commercially for at least one year.[216] Because inventors often sell their technology directly to larger companies, they will not meet the commercial practice requirement. Furthermore, very early stage start-ups will not have existed long enough to meet the one year requirement.[217] Accordingly, large companies will be advantaged, and have a competitive, low-cost benefit in the Act for which many small, fast-moving, innovative companies would not qualify. Even if true, it is not clear how this result would tend to undermine the performance of small companies, or to leave them in a position worse off than the one they held prior to the signing of the Act.

The director of a major federally funded lab echoed the assertion, reported elsewhere in this document, that "small innovative firms produce proportionately more, higher quality patents than large firms and they rely more heavily on patents to protect their innovations."[218] Claiming that the defense increases uncertainty over patent value and exclusivity,[219] it was suggested that prior use rights will have a disproportionate impact upon the economic performance of small businesses.

As a result of these differences of opinion, there did not appear to be agreement among the comments. As was the case in the university context, because prior user rights have only recently been made widely available, other respondents stated that there is insufficient evidence at this time to determine the impact on small business and innovators.[220]

Accordingly, after considering the available evidence, USPTO finds that:

Finding 6: There is no substantial evidence that the prior user right defense established by the AIA will have a negative impact on small businesses or independent inventors.

On the basis of the above analysis and findings, the USPTO recommends:

Recommendation 2: The prior user right defense under the AIA should be maintained with no change at the present time because there is no substantial evidence that it will have a negative impact on innovation, venture funding, small businesses, universities, or independent inventors.

Recommendation 3: The USPTO should reevaluate the economic impacts of prior user rights as part of its 2015 report to Congress on the implementation of the AIA, when better evidence as to these impacts might be available.[221]

C. Analysis of Legal and Constitutional Issues with Placing Trade Secret Law in Patent Law

The patent grant is derived from the Constitution, which authorizes Congress to secure to inventors the exclusive right to their discoveries.[222] Patent protection encourages individuals to apply their creative genius to producing new inventions by rewarding inventors exclusive control over their inventions for a limited time. In exchange for these exclusive rights, the patent holder is required to disclose to the world how that invention is made and used.[223] Because a prior use defense is viewed as an exception to the right of an inventor to exclude others from making and using the claimed invention, some have questioned the legal basis for such a defense. This section will offer an analysis of the legality of a prior use defense in United States patent law along with an analysis of the relationship between trade secret law and patent law.

In response to the Federal Register Notice requesting comments on the operation and impact of the prior user rights defense on patent infringement, the vast majority of respondents supported a prior use defense in United States patent law and did not address the legality of the prior use provision in the AIA. Of those who did address the legality of the provision, the majority did not express any concerns.[224]

In the majority view, there are no constitutional issues associated with adopting a prior user rights defense. The majority asserted that the Supreme

Court previously addressed the compatibility of trade secret and patent laws in *Kewanee Oil Co. v. Bicron Corp.*, sanctioning the co-existence of two.[225] The majority also asserted that the prior use defense has existed in the United States prior to the 1952 Patent Act, and was not deemed unconstitutional or even challenged based on constitutionality.[226] Additionally, some companies indicated that they should not be forced to disclose discoveries, which, even if patented, could be copied by others.[227]

A minority of respondents asserted that the prior user rights defense may be unconstitutional.[228] According to the minority view, the prior user right defense impedes the goal of the Constitution, "to promote the progress of science and the useful arts"[229] by encouraging inventors to keep their ideas secret.[230] Further, the minority also raised concerns about protecting trade secrets at the expense of patent rights.[231]

Analysis and Findings

1. Constitutional Analysis of Prior User Rights

Prior user rights are not unconstitutional. Instead, they comport with the Patent Clause of the Constitution, are consistent with the Patent Statute, and are supported by public policy considerations. Each is discussed, in turn. The Patent Clause[232] was "written against the backdrop" of the abuses of monopoly power; thus, Congress's ability to grant patents was intended to be limited.[233] The Patent Clause is aimed at rewarding exclusivity on those who create new inventions, but it does not prevent protecting prior users from liability for infringement to one who elects to secure a later patent. The Patent Clause is silent regarding prior users.

Accordingly, after the framers established the Patent Clause, later Congresses created prior user rights in the United States. The prior user right defense has existed in United States patent laws for over a hundred years, between 1839 and 1952 and then again after 1999.[234] Section 7 of the 1839 Patent Act stated that any person who constructed or purchased a new invention prior to the patent application for that invention had the right to use and sell that invention without liability to the patent holder.[235] This provision was revised from time-to-time and was ultimately repealed in 1952 as "unnecessary and redundant."[236] In 1999, in the wake of a Federal Circuit decision that confirmed the patentability of business methods,[237] a prior use defense was added solely to protect prior users of business methods.[238]

Further, Congress has imposed numerous other limits on a patent holder's monopoly, such as limiting the term of the patent monopoly, granting the power of eminent domain over patents, imposing limitations to prevent patent misuse, and restricting the scope of reissued patents, to name a few. The prior user right defense is another limitation on patent rights.

Finally, policy considerations support a prior user rights defense. In analyzing the constitutionality of the prior use defense it is imperative to evaluate whether such a defense would be considered a “rational exercise” of Congress’s legislative authority.[239] Testimony in the Congressional Record suggests that the prior use defense was intended to protect jobs by allowing companies who invented a manufacturing process to continue using that process.[240] Additionally, there is testimony that the prior use defense would relieve manufacturers of the burden of seeking patent protection for every manufacturing process they use, which could number in the hundreds or thousands.[241] There is also testimony that companies should not be forced to seek patent protection if the risks of copying due to disclosure in a patent outweigh the benefits of United States patent protection.[242] Finally, there is testimony that a prior use defense would put the United States on equal footing with foreign countries, most of whom have a prior use defense.[243]

Thus, the USPTO finds that:

Finding 7: A prior use defense to patent infringement is Constitutional and lawful and the defense is consistent with the Constitution and Supreme Court precedent recognizing that trade secret law and patent law can and do legally co-exist in the United States, and indeed have co-existed since our Constitution was created.

2.The Relationship between Trade Secret Protection and Patent Law

Somewhat related to the issue of the legality of the interplay between trade secret protection and patent law is whether a prior use defense to patent litigation strikes the appropriate balance between the two forms of intellectual property protection. One of the goals of the patent system is to enrich the fund of human knowledge through the timely disclosure of new technologies; trade secrets, by their very nature, do not accomplish this goal.[244] Some of the comments contend that providing prior user rights thwarts the aims of the patent system by favoring secrecy over disclosure.[245] The comments further argued that the grant of “co-exclusive rights” between the prior user and the patentee erodes the value of patents.[246] The problem, however, as pointed out in many of the comments, is that the patent system may not provide the right economic or strategic fit for each class of innovator or for each type of invention.[247]

Several of the comments noted that it is not economically feasible for certain businesses, particularly small- and medium-sized enterprises, to seek patent protection for every innovation they make.[248] For this class of innovators, a trade secret may be a cheaper alternative to a patent, one that provides a simpler means of protection and policing of a new technology.[249] By forgoing patent protection in favor of a trade secret, these entities can more efficiently allocate scarce resources to growth of other parts of the enterprise.[250]

Other comments noted that in certain industries, innovation tends to be incremental rather than pioneering, such as in the electronics and communications areas.[251] These comments pointed out that in such cases, the more economically rational and pro-competitive solution may be to protect advances as trade secrets, rather than through patent protection.[252] Moreover, forcing such minor advances into the patent system could, as observed by some comments, create a host of other problems, such as taxing USPTO resources and flooding the patent system with patents of dubious quality.[253]

As to the relationship between patents and trade secrets, the Supreme Court in *Kewanee Oil v. Bicorn Corp.* [254] recognized that trade secrets and patents can legally co-exist and that both trade secrets and patents have value. Additionally, in *Kewanee*, the Supreme Court recognized that trade secrets will not impede scientific progress (the goal of Article I, Section 8, Clause 8) because the ideas protected by trade secrets will likely be discovered by others in a similar time frame, stating that "[n]or does society face much risk that scientific or technological progress will be impeded by the rare inventor with a patentable invention who chooses trade secret protection over patent protection [...] [I]f a particular person had not made a discovery others would have, and in probably a relatively short period of time.[255] In *Kewanee*, the Supreme Court also noted that trade secrets can be put into public use and thereby disclosed to those who can independently discover the idea.[256]

A related issue is the consequence that, by filing a patent application, the technology is disclosed to competitors in other countries where the innovator may not, for lack of resources, have sought patent protection, or where patents are difficult to enforce.[257] Several comments pointed out that this can have drastic effects on American businesses and jobs. A representative from a small United States company commented that the company's use of trade secret protection enables it to provide services at a price level that its Chinese competitors admitted they cannot match.[258] Several representatives of "green" technology companies based in the United States noted that disclosure of new technologies via a patent application would allow foreign competitors to copy the innovation, using cheaper labor and with the assistance of foreign government subsidies, to undercut their businesses, at the expense of "creating much-needed jobs for American workers."[259]

Thus, the USPTO finds that:

Finding 8: Trade secret protection is of considerable value to United States businesses and the United States economy, and as such, there are compelling economic and policy justifications for providing a prior user rights defense to patent infringement.

On the basis of the above analysis and findings, the USPTO recommends as follows:

Recommendation 4: United States patent law should provide for a prior user rights defense as an appropriate balance between trade secret protection and patent protection, which legally co-exist to provide competitive advantages for United States businesses.

D. Analysis of Whether a First-to-File Patent System Creates a Need for Prior User Rights

One of the most fundamental changes under the AIA is in the way that entitlement to a patent is determined. The United States has historically awarded a patent to the first person to have invented the subject matter, commonly referred to as the "first-to-invent" standard. According to this approach, if two inventors file patent applications for the same invention, the patent is awarded to the first person to have conceived of the invention and to have reduced it to practice.[260]

Since 1998, the United States has been the only country in the world to use the first-to-invent standard.[261] Every other patent-granting country awards a patent to the first inventor to have filed an application for the subject matter. This approach, known as "first-to-file," relies on application filing dates, rather than dates of invention and reduction to practice, to determine priority as between competing applicants. Hence, the entity who files first has the right to the patent.[262] Section 3 of the AIA transitions the United States to a first-inventor-to-file patent system from a first-to-invent system and establishes derivation proceedings in place of interference proceedings.

In transitioning from a first-to-invent to first-inventor-to-file system, the AIA makes a number of ancillary changes to United States patent law, either as a direct result of, or consequential to, implementation of the first-inventor-to-file standard. One significant amendment in this respect is the expansion of the prior user rights defense set forth in the newly enacted section 273 of title 35.[263]

The switch from first-to-invent to first-inventor-to-file is cited in the legislative history of the AIA as the main impetus behind the changes to section 273.[264] One of the direct results of adopting a first-inventor-to-file regime is the elimination of section 102(g)(2) of the current law.[265] Section 102(g)(2) is the basis for the first-to-invent principle in current law and provides that an applicant is entitled to a patent unless "before the applicant's invention thereof, the invention was made . . . by another who had not abandoned, suppressed, or concealed it."[266] Under this provision, a prior user who did not abandon, suppress, or conceal the invention can invalidate the later patent of another and thereby escape liability for infringement.[267]

In a first-inventor-to-file regime, earlier invention is not a defense. As a result, the patent may be given to the first inventor to file an application for the invention. This raises the possibility, in the absence of a defense mechanism analogous to section 102(g)(2), that a party that earlier invented and commercially used the subject matter without disclosing it could be liable for infringement as against another party that later obtained a patent on it.

Analysis and Findings

Because the intersection of the first-inventor-to-file standard and prior user rights has been the subject of over two decades of legislative and academic debate in the United States, there is a wealth of commentary on the issue.[268] Most recently, in response to the Federal Register Notice requesting comments on the operation of prior user rights, the USPTO received twenty-nine comments, including both written submissions and oral testimony, which addressed the interaction between the first-inventorto-file standard and prior user rights. All of these comments expressed views about including prior user rights in a first-inventor-to-file system, some went even further and explicitly noted that prior user rights were either "essential," "critical," or otherwise necessary in such a system to balance various interests. Seven of the comments received, including from representatives of technology transfer offices or other entities involved in commercialization of upstream research, expressed negative views about prior user rights as a general matter, without specifying linkage to first-inventor-to-file. Some of the comments merely expressed skepticism or concern; others argued that prior user rights should not be permitted for a variety of reasons.

Despite the diversity of viewpoints expressed, the comments reveal a common analytical framework for considering the issue whether a first-inventor-to-file system creates a need for prior user rights. All of the responsive comments, like much of the extant literature on the subject, contain arguments either for or against prior user rights that can fairly be characterized as falling under one of the following headings: (i) fairness to the party who has expended effort and resources in using the later patented technology; and (ii) harmonization of patent systems.

1. Balancing Interests of the Prior User and the Patentee

The reason most frequently cited in support of prior user rights in a first-inventor-to-file regime is ensuring "fairness" -- appropriately balancing the equitable interests of the prior user and the patentee.[269] As noted above, a first-inventor-to-file system creates the potential for inequity if an earlier inventor commercializes a new technology without disclosing it, but another later obtains a patent on it. In such a situation, the earlier inventor may be subject to liability for infringement, which could entail a substantial loss of investment, loss of jobs, and erosion of U.S manufacturing. A prior user rights defense addresses this issue by allowing the earlier inventor to continue using the invention without liability, subject to certain conditions, while the patentee still enjoys exclusive rights as against everyone except the prior user.[270]

"Fairness" has been a common theme in the history of United States patent law.[271] Prior user rights have been specifically provided in the United States in the past as a means to balance respective interests.[272] The prior use defense applicable only to business methods, for instance, was enacted to protect businesses against infringement claims for using processes thought to be unpatentable before the decision in *State Street*.[273] Moreover, since 1952, United States patent law has provided a prior invention defense by virtue of section 102(g)(2), the elimination of which under the AIA creates a need for a similar defense applicable in a first-inventor-to-file environment to ensure a continued and similar balancing of interests under the new law.[274]

This same notion of "fairness" also underlies the provision of prior user rights in other jurisdictions.[275] A number of comments received were from, or on behalf of, foreign practitioners having experience with, or knowledge of, prior user rights in their home countries, *e.g.*, Japan, Germany and Australia. Several of these comments highlight the same basic dilemma created by a first-inventor-to-file regime, *i.e.*, that of an earlier commercial user and a later patentee of the same subject matter, as the motivating factor behind the establishment of prior user rights in those countries.[276] One German practitioner even noted that first-inventor-to-file is the "source" of prior user rights and that the potential inequity it causes makes prior user rights "essential" in such a system.[277]

Unfortunately, there is very little empirical data on the frequency of use of the prior user rights upon which to base a conclusion that providing them in a first-inventor-to-file regime is quantitatively essential. Many of the comments and the few studies on the subject note that there have been only a relative handful of litigation outcomes involving the prior user rights defense, either in the United States or abroad.[278] Some of the comments, however, explain that litigation outcomes only tell part of the story. The prior user rights defense frequently leads to pre-litigation settlements between the prior user and the patentee that are not – and in fact, could not be – revealed in empirical studies given the confidential nature of the settlements, or because the lack of pre-trial discovery in some jurisdictions prevents uncovering prior uses, limiting application of the defense.[279]

Thus, the USPTO finds that:

Finding 9: Providing limited prior user rights in a first-inventor-to-file system addresses the inherent inequity such a system creates between an earlier commercial user of the subject matter and a later patentee. A prior user rights defense is pro-manufacturing and pro-jobs, as it rewards businesses that put new technology promptly into commercial use, and provides protection for early commercial use when challenged by the later filing of patent applications by other entities.

2. Harmonization of Patent Laws

Several of the comments pointed out that the lack of a prior user rights regime in a first-inventorto-file system in the United States would put United States businesses at a disadvantage in comparison with its competitive trading partners, such as Europe, Canada, Japan, Korea, and Australia. Because these countries have prior use protections in their patent laws, this disharmony could disadvantage United States enterprises by making them liable for infringement in the United States as against a foreign-based United States patent holder, but unable to assert a foreign patent against a foreign entity in a jurisdiction that provides prior user rights.[280] Some comments further stated that this imbalance could lead to strategic decisions to locate businesses or components thereof, and the jobs that go along with them, outside the United States to prevent disruptions to operations or interference with the use of assets.[281]

Previous studies suggest that the magnitude of the potential imbalance is significant. One comment referenced a 1994 study that found that foreign countries representing about 85% of GDP outside the United States provided prior user rights.[282] Another study pointed out that United States interests were particularly at risk given the high percentage of United States patents awarded to foreign inventors.[283]

Thus, the USPTO finds that:

Finding 10: Because the availability of a prior user rights defense to patent infringement is a fundamental aspect of many patent regimes throughout the world, there is a strong preference that United States businesses be afforded the same advantages in terms of prior use protections in the United States that their competitors enjoy abroad. On the basis of the above analysis and findings, the USPTO recommends as follows:

Recommendation 5: United States patent law should provide for a prior user rights defense to patent infringement in order to address a systemic inequity inherent in a first-inventor-to-file system and to ensure United States businesses are

(1) able to protect their investments in the event of a later issued patent, and
(2) placed on similar footing as competitors in other jurisdictions.

IV. CONCLUSION

The availability of a prior use defense for alleged patent infringement is a fundamental aspect of many patent regimes throughout the industrialized world and the characteristics of the prior use defense in these countries are fairly consistent from country-to-country. The policy reasons for adopting a prior use defense are also fairly universal – prior user rights remedy the unfairness in holding a prior commercial user liable for patent infringement.

The recently enacted AIA incorporates a carefully crafted prior use defense consistent with the prior use defense found in many industrialized countries of the world, with certain unique features so that the only parties who can prove commercial use at least one year prior to the filing date of the patent application can obtain the defense. The scope of the defense is limited in the type activities that may be continued in relation to the original prior use activities so that the patentee's rights are not unjustly impinged. The prior user rights defense as set forth in the AIA is narrowly tailored and not expected to be asserted frequently in patent litigation. There is no substantial evidence that prior user rights will negatively impact innovation, start-up enterprises, venture capital, small businesses, universities or individual inventors. The USPTO should, however, reevaluate the economic impacts of prior user rights as part of its 2015 report to Congress on the implementation of the AIA when better evidence as to these impacts might be available.

A prior use defense to patent infringement, and specifically the one set forth in the AIA, is neither unconstitutional nor unlawful, as the defense is consistent with the Constitution and Supreme Court precedent recognizing that trade secret law and patent law can and do legally co-exist in the United States as they have for hundreds of years. Trade secret protection is of considerable value to United States businesses and the United States economy, and as such, there are compelling economic and policy justifications for providing a prior user rights defense to patent infringement. Providing suitably limited prior user rights in a first-inventor-to-file system is an appropriate response to an inherent inequity such a system creates as between an earlier commercial user of the subject matter and a later patentee. There is a strong preference that United States businesses be afforded the same advantages in terms of prior use protections in the United States that their competitors enjoy abroad.

APPENDIX A: PRIOR USE DEFENSE IN THE AMERICAN INVENTORS PROTECTION ACT OF 1999[284]

Defense to Infringement Based on Earlier Inventor

(a) DEFINITIONS— For purposes of this section—

(1) the terms "commercially used" and "commercial use" mean use of a method in the United States, so long as such use is in connection with an internal commercial use or an actual arm's-length sale or other arm's-length commercial transfer of a useful end result, whether or not the subject matter at issue is accessible to or otherwise known to the public, except that the subject matter for which commercial marketing or use is subject to a premarketing regulatory review period during which the safety or efficacy of the subject matter is established, including any period specified in section 156(g), shall be

deemed "commercially used" and in "commercial use" during such regulatory review period;

(2) in the case of activities performed by a nonprofit research laboratory, or nonprofit entity such as a university, research center, or hospital, a use for which the public is the intended beneficiary shall be considered to be a use described in paragraph (1), except that the use—

(A) may be asserted as a defense under this section only for continued use by and in the laboratory or nonprofit entity; and

(B) may not be asserted as a defense with respect to any subsequent commercialization or use outside such laboratory or nonprofit entity;

(3) the term "method" means a method of doing or conducting business; and

(4) the "effective filing date" of a patent is the earlier of the actual filing date of the application for the patent or the filing date of any earlier United States, foreign, or international application to which the subject matter at issue is entitled under section 119, 120, or 365 of this title.

(b) DEFENSE TO INFRINGEMENT—

(1) IN GENERAL— It shall be a defense to an action for infringement under section 271 of this title with respect to any subject matter that would otherwise infringe one or more claims for a method in the patent being asserted against a person, if such person had, acting in good faith, actually reduced the subject matter to practice at least 1 year before the effective filing date of such patent, and commercially used the subject matter before the effective filing date of such patent.

[...]

(2) LIMITATIONS AND QUALIFICATIONS OF DEFENSE— The defense to infringement under this section is subject to the following:

(A) PATENT— A person may not assert the defense under this section unless the invention for which the defense is asserted is for a method.

(B) DERIVATION— A person may not assert the defense under this section if the subject matter on which the defense is based was derived from the patentee or persons in privity with the patentee.

(C) NOT A GENERAL LICENSE— The defense asserted by a person under this section is not a general license under all claims of the patent at issue, but extends only to the specific subject matter claimed in the patent with respect to which the person can assert a defense under this chapter, except that the defense shall also extend to variations in the quantity or volume of use of the claimed subject matter, and to improvements in the claimed subject matter that do not infringe additional specifically claimed subject matter of the patent.

[...]

(6) PERSONAL DEFENSE— The defense under this section may be asserted only by the person who performed the acts necessary to establish the defense and, except for any transfer to the patent owner, the right to assert the defense shall not be licensed or assigned or transferred to another person except as an ancillary and subordinate part of a good faith assignment or transfer for other reasons of the entire enterprise or line of business to which the defense relates.

APPENDIX B: PRIOR USE DEFENSE IN THE LEAHY-SMITH AMERICA INVENTS ACT[285]

Defense to Infringement Based on Prior Commercial Use

(a) IN GENERAL—A person shall be entitled to a defense under section 282(b) with respect to subject matter consisting of a process, or consisting of a machine, manufacture, or composition of matter used in a manufacturing or other commercial process, that would otherwise infringe a claimed invention being asserted against the person if—

(1) such person, acting in good faith, commercially used the subject matter in the United States, either in connection with an internal commercial use or an actual arm's length sale or other arm's length commercial transfer of a useful end result of such commercial use; and

(2) such commercial use occurred at least 1 year before the earlier of either—

(A) the effective filing date of the claimed invention; or

(B) the date on which the claimed invention was disclosed to the public in a manner that qualified for the exception from prior art under section 102(b).

[...]

(e) LIMITATIONS AND EXCEPTIONS— (1) PERSONAL DEFENSE—

(A) IN GENERAL—A defense under this section may be asserted only by the person who performed or directed the performance of the commercial use described in subsection (a), or by an entity that controls, is controlled by, or is under common control with such person.

(B) TRANSFER OF RIGHT—Except for any transfer to the patent owner, the right to assert a defense under this section shall not be licensed or assigned or transferred to another person except as an ancillary and subordinate part of a good-faith assignment or transfer for other reasons of the entire enterprise or line of business to which the defense relates.

(C) RESTRICTION ON SITES—A defense under this section, when acquired by a person as part of an assignment or transfer described in subparagraph (B), may only be asserted for uses at sites where the subject matter that would otherwise infringe a claimed invention is in use before the later of the effective filing date of the claimed invention or the date of the assignment or transfer of such enterprise or line of business.

(2) DERIVATION—A person may not assert a defense under this section if the subject matter on which the defense is based was derived from the patentee or persons in privity with the patentee.

(3) NOT A GENERAL LICENSE—The defense asserted by a person under this section is not a general license under all claims of the patent at issue, but extends only to the specific subject matter for which it has been established that a commercial use that qualifies under this section occurred, except that the defense shall also extend to variations in the quantity or volume of use of the claimed subject matter, and to improvements in the claimed subject matter that do not infringe additional specifically claimed subject matter of the patent.

[...]

(5) UNIVERSITY EXCEPTION—

(A) IN GENERAL—A person commercially using subject matter to which subsection (a) applies may not assert a defense under this section if the claimed invention with respect to which the defense is asserted was, at the time the invention was made, owned or subject to an obligation of assignment to either an institution of higher education (as defined in section 101(a) of the Higher Education Act of 1965 (20 U.S.C. 1001(a)), or a technology transfer organization whose primary purpose is to facilitate the commercialization of technologies developed by one or more such institutions of higher education.

(B) EXCEPTION—Subparagraph (A) shall not apply if any of the activities required to reduce to practice the subject matter of the claimed invention could not have been undertaken using funds provided by the Federal Government.

APPENDIX C: LIST OF PARTIES PROVIDING TESTIMONY AND/OR COMMENTS[286]

Individuals Providing Testimony

Gary Griswold (on his own behalf)
Alan Kasper (AIPLA)
Thomas Kurkowski (on his own behalf)
Dan Lang (Cisco)
MaCharri Vorndran-Jones (ABA-IP Group)

Intellectual Property Organizations

America Bar Association – IP Section
American Intellectual Property Association (AIPLA)
Coalition for Patent Fairness
Computer & Communications Industry Association (CCIA)
Institute of Patent and Trade Mark Attorneys of Australia
International Federation of Intellectual Property Attorneys (FICPI)
Japan Intellectual Property Association Innovation Alliance
Japan Patent Attorneys Association
New Zealand Institute of Patent Attorneys

Academic and Research Institutions

Wisconsin Alumni Research Foundation
Berkeley Center for Law and Technology and Greentech Industries[287]
Regents of University of California Board
Higher Education Associations[288]

Companies

Belz Consulting
Cisco
Google and Verizon (Joint Comments)
Hospira, Inc.
Lex Machina
Microsoft Corp.
Space Exploration Technologies Corp.
Telstra Corporation

Individuals

Stephan Freischem
Asamichi Kato
Thomas E. Kelley
Thomas Kowalski
Paul Morgan
Tony Tether
Neil Thomas

APPENDIX D: COMPARISON CHART OF STUDIED COUNTRIES

	Prior Use Activities	Definition of Prior User	Scope of Prior User Activities	Transfer
Australia	Exploitation or definitive steps toward exploitation	Acquisition ok Requires good faith	Scope may not be extended from prior use	Freely assignable
Brazil	Exploitation of the patented subject matter	Acquisition ok Requires good faith	In same manner under same conditions	Only together with business, or part of business to which it belongs
Canada	Purchase, construction or acquisition of a later patented invention	Acquisition ok Requires good faith	Scope may not be extended from prior use	Encompasses purchase, construction or acquisition – transfer unclear
China	Use/manufacture or making necessary preparations for same	Acquisition ok Requires good faith	Scope may not be extended from prior use	Only together with business
Denmark	Exploiting the invention commercially or substantial preparations	Acquisition ok Requires good faith	Prior User Right exploitation must retain its general character	Only together with business
France	Possession of the invention (preparations of an industrial/productive nature)	Acquisition ok Requires good faith	Any equivalents of the patented invention	Only together with business, or part of business to which it belongs

Appendix D: (Continued)

	Prior Use Activities	Definition of Prior User	Scope of Prior User Activities	Transfer
Germany	Use or necessary arrangements to do the same	no	Limited to the particular needs of the business	Only together with business
Japan	Working the invention or preparing for the same	Acquisition ok Requires good faith	Extent of invention on purpose of such business	Only together with business, or with patent owner's consent or by succession
Republic of Korea	Working the invention or preparing for the same	Acquisition ok Requires good faith	Scope of the objective of the invention or the related business	Only together with business or with patent owner's consent or by succession
Mexico	Use or necessary preparations for the same	Acquisition ok No good faith requirement in law	Scope may not be extended from prior use	Prior user rights are not transferable
Russian Federation	Use or necessary preparations for the same	Requires conception; requires good faith	Scope may not be extended from prior use	Only together with business, or part of business to which it belongs
United Kingdom	Acts that would constitute infringement or effective and serious preparation	Acquisition ok Requires good faith	Does not affect the essence of the invention	Only together with business, or part of business to which it belongs
United States	One full year of commercial use	Requires reduction to practice Requires good faith	Scope that might otherwise be considered infringing use	Only together with the entire enterprise or line of business

End Notes

[1] Leahy-Smith America Invents Act, Pub. L. No. 112-29, § 3(m), 125 Stat. 284, 292 (2011).

[2] *Id.*

[3] Leahy-Smith America Invents Act, Pub. L. No. 112-29, 125 Stat. 284 (2011).

[4] H.R. REP. NO. 112-98, at 38 (2011).

[5] U.S. CONST. art. I, § 8, cl. 8.

[6] *See* H.R. REP. NO. 112-98, at 38-40.

[7] *Id.* at 40.

[8] The terms "prior user rights" and "prior use defense" are used interchangeably in this Report.

[9] *See generally* Keith Kupferschmid, *Prior User Rights: The Inventor's Lottery Ticket*, 21 AIPLA Q.J. 213, 216 (1993) (defining detailed requirements for satisfying the defense), *and* Gary L. Griswold & F. Andrew Ubel, *Prior User Right—A Necessary Part of a First-to-File System*, 26 J. MARSHALL L. REV. 567, 572 (1993).

[10] *See, e.g.,* H.R. REP. NO. 112-98, at 44 (noting, in particular, the tendency of high-technology businesses not to file for patent protection on every innovation).

[11] *See* Kupferschmid, *supra* note 9, at 217, 229.

[12] *Id.*

[13] H.R. REP. NO. 112-98, at 40.

[14] *See* discussion *infra* Part III.C.

[15] *See* Patent Act of 1836, ch. 357, § 6, 5 Stat. 117 (1836); Patent Act of 1839, ch. 88, § 7, 5 Stat 353 (1839); *and* Patent Act of 1870, ch. 230, § 24, 16 Stat. 198 (1870). Some commentators further argue that existing U.S. patent law continues to provide for a prior user rights defense even in the absence of an explicit provision. *See, e.g.*, Griswold & Ubel, *supra* note 9, at 591 and 572 n.20 (quoting Judge Pauline Newman of the U.S. Court of Appeals for the Federal Circuit as saying "I have not seen anyone who was a prior user who has been stopped upon raising the [35 U.S.C. §] 102(g) defense and from that viewpoint, it seems that the prior user right is alive and well . . . [I]f we stay with the current first-to-invent system, we would be changing direction if we felt that there should not be prior user right."); *accord* Karl F. Jorda, *Patent and Trade Secret Complimentariness: An Unsuspected Synergy*, 48 WASHBURN L.J. 1, 27 (2008) (pointing out that there has never been a reported case in the United States where a patentee has enjoined a prior user/trade secret owner).

[16] 35 U.S.C. § 273 (2006). *See* App. A (section 273 of the AIPA is reproduced for reference).

[17] 149 F.3d 1368 (Fed. Cir. 1998), *cert. denied*, 525 U.S. 1093 (1999).

[18] *Id.* at 1375.

[19] H.R. Rep. 106-464, at 122 (1999) ("The 1998 opinion by the U.S. Court of Appeals for the Federal Circuit in *State Street Bank and Trust Co. v. Signature Financial Group*, which held that methods of doing business are patentable, has added to the urgency" of providing prior user rights vis-à-vis such patents.).

[20] ADVISORY COMM.'N ON PATENT LAW REFORM, 102ND CONG., REPORT TO THE SEC. OF COM. (Comm. Print 1992); Patent Harmonization Act of 1992, H.R. 4978, 102d Cong. § 3 (1992); Patent Harmonization Act of 1992, S. 2605, 102d Cong. § 3 (1992); Prior User Rights Act of 1994, S. 2272, 103d Cong. (1994); Prior Domestic Commercial Use Act of 1995, H.R. 2235, 104th Cong. (1995); Inventor Rights Protection and Patent Reform Act of 1996, H.R. 3460, 104th Cong. §§ 301-303 (1996) (also known as the "Moorhead-Schroeder Patent Reform Act"); Omnibus Patent Act of 1996, S. 1961, 104th Cong. §§ 401-403 (1997); Omnibus Patent Act of 1997, S. 507, 105th Cong. §§ 401-403 (1997); 21st

Century Patent Improvement Act, H.R. 400, 105th Cong. §§ 301-303 (1997). *Compare* Robert L. Rohrback, *Prior User Rights: Roses or Thorns?*, 2 U. BALT. INTELL. PROP. L.J. 1, 11-28 (1993) *with* Gary L. Griswold et al., *Letter to the Editor: Prior User Rights: Neither A Rose Nor A Thorn*, 2 U. BALT. INTELL. PROP. L.J. 233, 233-34 (1994) (specifically rebutting Rohrback by arguing that prior user rights would more fully reconcile trade secret and patent law); *and* Frank E. Robbins, *The Rights of the First Inventor-Trade Secret User as Against Those of the Second Inventor-Patentee (Part I)*, 61 J. PAT. & TRADEMARK OFF. SOC'Y 574, 587-89 (1979) *with* Karl F. Jorda, *The Rights of the First Inventor-Trade Secret User as Against Those of the Second Inventor-Patentee (Part II)*, 61 J. PAT. & TRADEMARK OFF. SOC'Y 593, 604-05 (1979).

[21] § 5(a), 125 Stat. at 297. *See* App. B (section 273 under the AIA is reproduced for reference).

[22] *Id.*

[23] *Id.*

[24] *Id.* at 298.

[25] *Id.*

[26] *Id.* at 298-99.

[27] *See America Invents Act: Hearing on H.R. 1249 Before the Subcomm. on Intellectual. Property, Competition, and the Internet of the House Comm. on the Judiciary*, 112th Cong. 97-98 (2011) (Statement of John C. Vaughn, Exec. Vice President, Ass'n of Am. Univs.); *see also Prior User Rights (Relative to Patents): Hearing Before the Subcomm. on Intellectual Property and Judicial Administration of the House Comm. on the Judiciary*, 103d Cong. 99-101 (1995) (Statement of Teri F. Willey, Purdue Research Found.).

[28] § 3(m), 125 Stat. at 292.

[29] Notice of Public Hearing and Request for Comments on the Study of Prior User Rights, 76 Fed. Reg. 62,388 (Oct. 7, 2011).

[30] *Id.* at 62,389.

[31] *Id.*

[32] *Id.*

[33] *Infra* App. C (list of parties providing testimony and/or comments); *see also* Comments on the Study of Prior User Rights, http://www.uspto.gov/aia_implementation/prior-user-rights-comments.jsp (last visited Dec. 7, 2011).

[34] § 3(m), 125 Stat. at 292.

[35] Most members of the European Union are also contracting members of the European Patent Convention. *See generally* European Patent Office, Member states of the European Patent Organisation, http://www.epo.org/aboutus/organisation/member-states.html (last visited December 16, 2011). However, the European Patent Convention (establishing the European Patent Office) is not a treaty under the European Union. Further, the European Patent Convention does not currently establish a unitary patent system.

[36] *Id.*

[37] *See generally* Comments from Telstra Corp., to USPTO, Prior User Rights (Nov. 8, 2011) [hereinafter "Comments of Telstra Corp."], http://www.uspto.gov/aia_implementation/pur-2011nov08-telstra_corp.pdf; *and* Comments from David Tadgell, Convenor for the International Patents Committee, Institute of Patent and Trade Mark Attorneys, to the USPTO, Comments on the Study of Prior User Rights (Nov. 10, 2011) [hereinafter "Comments of Inst. of Patent and Trade Mark Attorneys of Australia"], http://www.uspto.gov/aia_implementation/ pur-2011nov10-ipta.pdf.

[38] *See generally* Comments from Tim Jackson, President, New Zealand Inst. of Patent Attys., Inc., to David Kappos et al., Undersecretary Dept. of Comm. for Intell'l Prop., and Director

of the USPTO (Nov. 9, 2011) [hereinafter "Comments of New Zealand Inst. of Patent Attys., Inc."], http://www.uspto.gov/aia_implementation/pur2011nov09-nzipa_inc.pdf.

[39] Comments from Stephan Freischam, European Patent Attorney, Patentanwaelte Freischem, to Elizabeth Shaw, IP Research Specialist Supervisor, USPTO, Office of Policy and External Affairs, Prior User Rights (Nov. 8, 2011) [hereinafter "Comments of Freischem"], http://www.uspto.gov/aia_implementation/pur-2011- stephan_freischem.pdf.

[40] Comments from the International Federation of Intell. Prop. Attorneys, to the U.S. Patent and Trademark Office (Nov. 16, 2011) [hereinafter "Comments from FICPI"], http://www.uspto.gov/aia_implementation/pur-2011nov16- ficpi.pdf.

[41] Comments of Telstra Corp., *supra* note 37, at 1, *and* Comments of New Zealand Inst. of Patent Attys., Inc., *supra* note 38, at 1.

[42] Comments of Inst. of Patent and Trade Mark Attorneys of Australia, *supra* note 37, at 2; Comments of Freischem, *supra* note 39, at 2; *and* Comments of FICPI, *supra* note 40, at 2.

[43] Comments from Jason Albert, Associate General Counsel for Intell. Prop. Policy & Strategy, Microsoft Corp., to USPTO, Comments of Microsoft Corp. (Nov. 8, 2011) [hereinafter "Comments of Microsoft Corp."], http://www.uspto.gov/aia_implementation/pur-2011nov08-microsoft_corporation.pdf; *and* Comments from Leah Taylor, Vice-President for Intell'l Prop., Hospira Inc., to Elizabeth Shaw, IP Research Specialist Supervisor, USPTO, Office of Policy and External Affairs, Prior User Rights (Nov. 8, 2011) [hereinafter "Comments of Hospira, Inc."], http://www.uspto.gov/aia_implementation/pur-2011nov08-hospira_inc.pdf ("a global specialty pharmaceutical and medication delivery company").

[44] Comments from Brian Pomper, Executive Director, Innovation Alliance, to Elizabeth Shaw, IP Research Specialist Supervisor, USPTO, Office of Policy and External Affairs, Comments on the Study of Prior User Rights (Nov. 8, 2011) [hereinafter "Comments of Innovation Alliance"], http://www.uspto.gov/aia_implementation/pur2011nov08-hospira_inc.pdf.

[45] Comments of Microsoft Corp., *supra* note 43, at 2. *See generally* Comments of Hospira, Inc., *supra* note 43.

[46] Comments of Innovation Alliance, *supra* note 43, at 1.

[47] Comments from Robert Barr et al., Executive Director, Berkeley Center for Law and Tech., Univ. of Cal., Boalt Hall Law School, to David Kappos et al., Undersecretary Dept. of Comm. for Intell'l Prop. and Director of the USPTO, Prior User Rights for Venture Capital Backed Greentech Indus. 2-3 (Nov. 6, 2011) [hereinafter "Comments of Greentech"], http://www.uspto.gov/aia_implementation/pur-2011nov06-berkeley-center.pdf; Comments from Thomas Kelley, Consulting Patent Counsel, Monsanto Co., to the USPTO, Prior User Rights 2 (Nov. 8, 2011) [hereinafter "Comments of Thomas Kelly"], http://www.uspto.gov/aia_implementation/pur-2011nov08- thomas_kelley.pdf; Thomas Kowalski, Shareholder, Vedder Price P.C., Comments at the Public Hearing on the Study of Prior User Rights 19 (Oct. 25, 2011) [hereinafter "Testimony of Thomas Kowalski"], http://www.uspto.gov/aia_implementation/20111025-prior-user-rights-transcript.pdf; Comments from James F. Kurkowski, Chief Intell. Prop. Counsel, Space Exploration Techs. Corp., to the USPTO, Comments of Space Exploration Techs. Corp. 4-5 (Nov. 8, 2011) [hereinafter "Comments of Space Exploration Techs. Corp."] http://www.uspto.gov/aia_implementation/pur-2011nov08-space_exploration_tech_corp.pdf; Comments from Dan Lang, Vice President Intellectual Property for Cisco Systems, Inc., to the USPTO, Written Comment on behalf of the Coalition for Patent Fairness 4-7 (Nov. 8, 2011) [hereinafter "Comments from Coalition for Patent Fairness"] http://www.uspto.gov/aia_implementation/pur-2011oct25-cisco.pdf.

[48] Comments of Greentech, *supra* note 47, at 1-2; *and* Comments of Space Exploration Techs. Corp., *supra* note 47, at 2.

[49] *See generally* Comments of FICPI, *supra* note 40, at 3 (noting that in some jurisdictions "the prior use covers the mere intellectual possession of the same invention, independently from the intention to effectively exploit such invention. In other jurisdictions . . . the 'use' is restricted to an effective industrial use.").

[50] *See generally id.* ("Other jurisdictions have an in-between approach, where the use is extended to serious and effective preparation to such industrial use.").

[51] *See, e.g.*, Patents Act, 1990 (Austl.), § 119(5) [hereinafter "Australian Patent Law"] (defining exploitation differently based on the type of invention).

[52] *See* Kenneth Félix Haczysnki da Nóbrega, Head of the Intell'l Prop. Division, Ministry of External Relations, Submission in Response to WIPO Questionnaire on Exceptions and Limitations to Patent Rights, 9-11, http://www.wipo.int/export/sites/www/scp/en/exceptions/submissions/brazil.pdf (last visited Dec. 17, 2011); and Comments from FICPI, *supra* note 49, at 3.

[53] Law No. 9.279 of 14 May 1996, Sec. II, art. 45 (Brazil) [hereinafter "Brazilian Patent Law"] ("A person who in good faith, prior to the filing or priority date of a patent application, was exploiting the object thereof in this country, shall be assured the right to continue the exploitation, without onus, in the same manner and under the same conditions as before . . . .").

[54] Canada Patent Act, R.S.C., 1985, ch. P-4, § 56(1) (2006) [hereinafter "Canadian Patent Act"].

[55] Act No. 92-597 of July 1, 1992 of the Intell. Prop. C., Official Journal (France), July 3, 1992, art. L613-7, *amended by* Act No. 96-1106 of December 18, 1996 of the Intell. Prop. C., Art. 4, Official Journal, Dec. 19, 1996 (updates last consolidated by Act No. 2006-236 of Mar. 1, 2006 of the Intell. Prop. C., Official Journal, Mar. 2, 2006.) [hereinafter "French Intell'l Prop. Law"] ("Any person who, within the territory in which this Book applies, at the filing date or priority date of a patent was, in good faith, in possession of the invention which is the subject matter of the patent shall enjoy a personal right to work that invention despite the existence of the patent."); *see also* Testimony of Thomas Kowalski, *supra* note 47, at 19 (noting that in France "mere possession of an invention prior to the patent application can be sufficient to defeat an infringement claim").

[56] *Compare* Francois Pochart, *Prior User's Rights in France*, UNION-IP (European Practitioners in Intell. Prop.) Congress Port. 7-12 (2008), *with* Daphne DeBeco, Official Representative, International and European Affairs Service, National Institute of Indus. Prop. (Fr.), Responses to WIPO Questionnaire on Exceptions and Limitations to Patent Rights, http://www.wipo.int/export/sites/www/scp/en/exceptions/translation/france_en.pdf (last visited Dec. 16, 2011) (indicating that "the invention must be *known completely*.").

[57] *Id.*

[58] Law No. 91 of 28 January 2009, Consolidate Patents Act, §4(1) (Den.) [hereinafter "Danish Patents Act"] ("Any person who, at the time when the patent application was filed, was exploiting the invention commercially in this country may, notwithstanding the grant of a patent, continue such exploitation retaining its general character [...] Such a right of exploitation shall also, under similar conditions, be enjoyed by any person who had made substantial preparations for commercial exploitation of the invention in this country."); *see generally* Flemming K. Mejl, Chief Technical Adviser, Danish Patent and Trademark Office, Response to WIPO Questionnaire on Exceptions and Limitations to Patent Rights 8-9, http://www.wipo.int/export/sites/www/scp/en/exceptions/submissions/ denmark.pdf.

[59] Patent Act, Act No. 121 of 1959, art. 79 (amended by Act No. 75 of June 29, 2005.) [hereinafter "Japanese Patent Act"] (where use is defined as "working the invention", and possession is defined as making "preparing" for the same).

[60] Patent Act, Dec. 31, 1961, art. 103 (last amended by Act No. 9381, Jan. 30, 2009.) [hereinafter "Korean Patent Act"] (where use is defined as "working the invention", and possession is defined as making "preparations" for the same).

[61] Indus. Prop. Law, Mexican Federation Official Diary, Art. 22.III, January 25, 2006 [hereinafter "Mexican IP Law"] (where use is defined simply as use, and possession is defined as making "necessary preparations" for the same).

[62] Patent Law of the People's Republic of China, art. 69(2) (promulgated by the Standing Comm. Nat'l People's Cong., Dec. 27, 2008, effective Oct. 1, 2009.) [hereinafter "Chinese Patent Law"] (where use is defined as "use/manufacture," and possession is defined as "making necessary preparations" for the same).

[63] Patent Law, 2003 (Russ.), § IV, art. 12 [hereinafter "Russian Patent Law"] (where use is defined simply as use, and possession is defined as making "necessary preparations" for the same).

[64] Patents Act, 1977, c. 37, § 64 (Eng.) [hereinafter "U.K. Patent Act"] ("(1) Where a patent is granted for an invention, a person who in the United Kingdom before the priority date of the invention does in good faith an act which would constitute an infringement of the patent if it were in force, or makes in good faith effective and serious preparations to do such an act, shall have the rights conferred by subsection (2). . . .").

[65] Andrew Webb, *Prior Use Under Section 64 of the UK Patents Act 1977*, UNION-IP, white paper 4.6-4.10 (European Practitioners in Intell. Prop.) Congress Port. (2008) (citing *Helitune Ltd. v. Stewart Hughes Ltd.*, [1991] F.S.R. 171, and *Lubrizol Corp., et al. v. Esso Petroleoum Co., et al.*, [1998] R.P.C. 727 (CA)).

[66] Patent Act (F.R.G.), Dec. 16, 1980, BGBl. I at 1981, last amended by Gazette (F.R.G.), Jul. 31, 2009, BGBl. I at 2521, pt. 1, § 12 [hereinafter "German Patent Act"].

[67] Sabine Rojahn, *Right of Private Prior Use: Sec. 12 Patent Act*, UNION-IP (European Practitioners in Intell. Prop.) Congress Port., slides 5-8 (2008).

[68] *Id.*

[69] German Patent Act, at § 12(1); *see also* German Patent and Trademark Office, Response to WIPO Questionnaire on Exceptions and Limitations to Patent Rights 8, http://www.wipo.int/export/sites/www/scp/en/exceptions/ submissions/germany.pdf (last visited Dec. 21, 2011).

[70] Rojahn, *supra* note 67, at slide 10.

[71] *Id.*

[72] *Id.* at slide 17.

[73] § 5, 125 Stat. at 297 (to be codified at 35 U.S.C. § 273(a)).

[74] *Id.* (to be codified at 35 U.S.C. § 273(c)).

[75] *Id.* (to be codified at 35 U.S.C. § 273(c)(1)).

[76] *Id.* (to be codified at 35 U.S.C. § 273(c)(2)).

[77] Australian Patent Law, at § 119(5).

[78] *Id.*; *See also* Adam Wright, Assistant Director, International Policy and Cooperation, IP Australia, Response to WIPO Questionnaire on Exceptions and Limitations to Patent Rights 12-15, http://www.wipo.int/export/sites/www/scp/en/exceptions/submissions/australia.pdf (last visited Dec. 16, 2011), *and* Comments of Inst. of Patent and Trade Mark Attorneys of Australia, *supra* note 37.

[79] Australian Patent Law, at § 119(5).

[80] *Id.*

[81] Japanese Patent Act, at art. 2(3).

[82] Korean Patent Act, at art. 103.

[83] Of note is that prior use in Japan also encompasses "providing [computer software] through an electric telecommunication line." Japanese Patent Act art. 2(3); *see also* Yuichiro Nakaya & Gen Aida, Deputy Director & Assistant Director, Japan Patent Office, Response to WIPO Questionnaire on Exceptions and Limitations to Patent Rights 8-12, http://www.wipo.int/export/sites/www/scp/en/exceptions/submissions/japan_rev.pdf (last visited Jan. 13, 2012).

[84] § 5, 125 Stat. at 297 (to be codified at 35 U.S.C. § 273(a)).

[85] Russian Patent Law, at art. 12.

[86] *Id.*

[87] *See generally* U.K. Patent Act, at § 64; German Patent Act § 12; Danish Patents Act, at § 4(1); French Intell'l Prop. Law, at art. L613-7; Canadian Patent Act, at § 56(1); Brazilian Patent Law, at art. 45; Mexican IP Law, at Art. 22.III; Australian Patent Law, at § 119; Japanese Patent Act, at art. 79; Korean Patent Act, at art. 103; *and* Chinese Patent Law, at art. 69(2).

[88] U.K. Patent Act, at § 64.

[89] Webb, *supra* note 65, at 8.1.

[90] § 5, 125 Stat. at 298 (to be codified at 35 U.S.C. § 273(e)(1)(A)).

[91] *Id.* (to be codified at 35 U.S.C. § 273(a)).

[92] German Patent Act, at § 12(1) ("If the applicant or his predecessor in title, before applying for a patent, disclosed the invention to other persons and reserved his rights in the event of a patent being granted, said person learning of the invention as a result of such disclosure cannot invoke measures under the provisions of the first sentence, which he has taken within six months after the disclosure.").

[93] Rojahn, *supra* note 67, at slide 8.

[94] Rojahn, *supra* note 67, at slide 8.

[95] *Id.*

[96] Danish Patents Act, at § 4(1).

[97] Australian Patent Law, at § 119(3).

[98] Japanese Patent Act, at art. 79.

[99] Another narrowly defined good faith standard is that of Republic of Korea. Korean Patent Act, at art. 103 (good faith standard is defined as "a person without knowledge of the content of an invention").

[100] French Intell'l Prop. Law, at art. L613-7.

[101] Pochart, *supra* note 56, at 5.

[102] U.K. Patent Act, at §64(1).

[103] Brazilian Patent Law, at art. 45.

[104] Russian Patent Law, at art. 12.

[105] Gregor Binkley, *Prior User Rights in the Canadian Patent Act*, 18 C.I.P.R. 207, 226 (2001).

[106] Chinese Patent Law, at art. 69(2).

[107] Liu Xiaojun, *A Synopsis on Defenses to Patent Infringement*, China Intell. Prop., Mar. 6, 2011, *available at* http://ipr.chinadaily.com.cn/2011-03/06/content_12124863.htm. *See also* Judicial Interpretation of the Supreme People's Court on Several Issues concerning the Application of Law in the Trial of Patent Infringement Dispute Cases, art. 15 (promulgated by the Supreme Judicial Committee, Dece. 21, 2009, effective Jan. 1, 2010).

[108] *Id.*

[109] Mexican IP Law, at art. 22.III.

[110] § 5, 125 Stat. at 297 (to be codified at 35 U.S.C. § 273(a)).

[111] *Id.* (to be codified at 35 U.S.C. § 273(e)(2)).

[112] *See generally* U.K. Patent Act, at § 64; German Patent Act § 12; Danish Patents Act, at § 4(1); French Intell'l Prop. Law, at art. L613-7; Canadian Patent Act, at § 56(1); Brazilian Patent Law, at art. 45; Mexican IP Law, at Art. 22.III; Australian Patent Law, at § 119; Japanese Patent Act, at art. 79; Korean Patent Act, at art. 103; *and* Chinese Patent Law. at art. 69(2).

[113] U.K. Patent Act, at § 64.

[114] U.K. Patent Act, at § 5.

[115] German Patent Act, at § 12(2).

[116] *Id.*

[117] *Id.*

[118] § 5, 125 Stat. at 297 (to be codified at 35 U.S.C. § 273(a)).

[119] *See generally*: U.K. Patent Act, at § 64; German Patent Act § 12; Danish Patents Act, at § 4(1); French Intell'l Prop. Law, at art. L613-7; Canadian Patent Act, at § 56(1); Brazilian Patent Law, at art. 45; Mexican IP Law, at Art. 22.III; Australian Patent Law, at § 119; Japanese Patent Act, at art. 79; Korean Patent Act, at art. 103; *and* Chinese Patent Law, at art. 69(2).

[120] Japanese Patent Act art. 79.

[121] French Intell'l Prop. Law, at art. L613-7.

[122] Pochart, *supra* note 56, at slide 3.

[123] Webb, *supra* note 65 at 7.1-7.3 (noting that the territorial requirement in U.K. may contravene European Competition Law by discriminating between acts carried on in different states of the EU).

[124] § 5, 125 Stat. at 297 (to be codified at 35 U.S.C. § 273(a)).

[125] Law No. 91 of 28 January 2009, Consolidate Patents Act, §4(1) (Denmark); *See also* PATRICK VAN EECKE ET AL., MONITORING AND ANALYSIS OF TECH. TRANSFER AND INTELL. PROP. REGIMES AND THEIR USE: RESULTS OF A STUDY CARDED OUT ON BEHALF OF THE EUROPEAN COMMISSION 104-105 (DG Research 2009) (noting that Denmark explicitly limits the extent of the prior user rights).

[126] Law No. 9.279 of 14 May 1996, Sec. II, art. 45 (Brazil) [hereinafter "Brazilian Patent Law"] ("A person who in good faith, prior to the filing or priority date of a patent application, was exploiting the object thereof in this country, shall be assured the right to continue the exploitation, without onus, in the same manner and under the same conditions as before . . . .").

[127] Russian Patent Law, § IV, art. 12.

[128] Australian Patent Law at § 119(5).

[129] Canadian Patent Act at § 56(1) (articulating that prior users have "the right to use and sell to others the specific article, machine, manufacture or composition of matter *patented* and so purchased, constructed or acquired without being liable to the patentee or the legal representatives of the patentee for so doing").

[130] Mexican IP Law, at art. 22.III.

[131] Chinese Patent Law art. 69(2).

[132] Korean Patent Act at art. 103.

[133] Hiroo Maeda, *Seven New Frameworks for Protection and Management of Knowledge*, INST. INTELL. PROP. BULL. 50, 55 (2006), *available at* http://www.iip.or.jp/e/e_summary/pdf/detail2005/e17_07.pdf.

[134] Japanese Patent Act at art.79 (noting that the non-exclusive license is provided "only to the extent of the invention and the purpose of such business worked or prepared").

[135] U.K. Patent Act, at § 64.

[136] VAN EECKE, *supra* note 125, at 104 (noting that legal doctrine of the U.K. provides that a prior user may expand into other products).

[137] VAN EECKE, *supra* note 125, at 104.

[138] Pochart, *supra* note 56, at slide 12.

[139] Patent Act (Germany), Dec. 16, 1980, BGBl. I at 1981, last amended by Gazette (Germany), July 31, 2009, BGBl. I at 2521, pt. 1, § 12 [hereinafter "German Patent Act"].

[140] § 5, 125 Stat. at 298 (to be codified at 35 U.S.C. § 273(e)(3)).

[141] *Id.*

[142] Australian Patent Law, at § 119(4).

[143] IP Australia Submission, at 12-15.

[144] *Id.* at § 4(2).

[145] German Patent Act, at § 12.

[146] Chinese Patent Law, at art. 69(2).

[147] Japanese Patent Act, Article 94(1); *see also* Yuichiro Nakaya & Gen Aida, Deputy Director & Assistant Director, Japan Patent Office, Submission re WIPO Questionnaire on Exceptions and Limitations to Patent Rights, 8-9, http://www.wipo.int/export/sites/www/scp/en/exceptions/submissions/japan_rev.pdf (last visited Dec. 21, 2011).

[148] Maeda, *supra* note 133, at 55; *see also* Hyun-suk Lim, Senior Deputy Director, Korean Intell'l Prop. Office, Submission re Questionnaire on Exceptions and Limitations to Patent Rights, 8-9, http://www.wipo.int/export/sites/www/scp/en/exceptions/submissions/rep_of_korea.pdf (A prior user right in the Republic of Korea can be transferred together with the business, in the case of inheritance or other general succession, or with consent of the patent holder).

[149] French Intell'l Prop. Law, at art. L613-7.

[150] Russian Patent Law, at art. 12.

[151] Brazilian Patent Law at art. 45(2) (under Brazilian law prior user rights "may only be assigned together with the business or company, or part thereof that is directly related to the exploitation of the object of the patent, by transfer or leasing.")

[152] U.K. Patent Act, at § 64.

[153] VAN EECKE, *supra* note 125, at 105; *see also* Hazel Craven, Senior Legal Adviser, UK Intell'l Prop. Office, Submission re WIPO Questionnaire on Exceptions and Limitations to Patent Rights, 16-18, http://www.wipo.int/export/sites/www/scp/en/exceptions/submissions/uk.pdf; *and* Webb, *supra* note 65, at 6.

[154] Mexican IP Law, Art. 22.

[155] Fabian R. Salazar, Divisional Director for Patents, Mexican Institute for Indus. Prop., Submission re WIPO Questionnaire on Exceptions and Limitations to Patent Rights, 10, http://www.wipo.int/export/sites/www/scp/es/ exceptions/submissions/mexico.pdf.

[156] Canadian Patent Act at § 56(1).

[157] Binkley at 208.

[158] § 5, 125 Stat. at 298 (to be codified at 35 U.S.C. § 273(e)(1)(B)).

[159] *Id.*

[160] § 5, 125 Stat. at 298-299 (to be codified at 35 U.S.C. § 273(e)(5)).

[161] §§ 3(m)1.B.-C., 125 Stat. at 292.

[162] *See, e.g.*, Kenneth J. Arrow, *Economic Welfare and the Allocation of Resources for Inventions*, *in* THE RATE AND DIRECTION OF INVENTIVE ACTIVITY: ECONOMIC AND SOCIAL FACTORS 609-625 (Nat'l Bureau of Econ. Research 1962).

[163] *Id.*

[164] Comments from William G. Barber, President, Am. Intell'l Prop. Ass'n, to Elizabeth Shaw, IP Research Specialist Supervisor, USPTO, Office of Policy and External Affairs, Prior User Rights 4 (Nov. 8, 2011) [hereinafter "Comments of AIPLA"], http://www.uspto.gov/aia_implementation/pur-2011nov08-aipla.pdf; Comments of FICPI, *supra* note 40, at 4. The analysis considers that the impact of such a defense will tend to be roughly similar across nations that share similar innovation systems, economic structures, and technological distributions. Because the OECD nations tend to share these same general characteristics, these impacts are addressed in general terms, not on a country-by-country basis.

[165] *See generally* Comments of Microsoft Corp., *supra* note 43, at 6.

[166] *See generally* Wesley M. Cohen et al., *Protecting Their Intellectual Assets: Appropriability Conditions and Why U.S. Manufacturing Firms Patent (or Not)* (Nat'l Bureau of Econ. Research, Working Paper No. 7552 (2000)), *available at* http://www.nber.org/papers/w7552.

[167] 35 U.S.C. § 273.

[168] *Seal-Flex, Inc. v. W.R. Dougherty and Assocs.*, 179 F. Supp. 2d 735, 741-42 (E.D. Mich. 2002) (refusing to address the prior user defense on the basis that it was untimely raised). Some cases covered § 273 in an ancillary manner. *See generally Bilski v. Kappos*, 130 S. Ct. 3218, 3228-29 (2010) (pointing out that the existence of section 273 suggests that business methods are not categorically excluded from patent-eligibility); *Aristocrat Techs. Austl. PTY Ltd. v. Int'l Game Tech.*, 543 F.3d 657, 662 (Fed. Cir. 2008) (dicta on § 273); *PB Farradyne, Inc. v. Peterson*, 2006 U.S. Dist. LEXIS 67281, *12 (N.D. Cal. 2006) (plaintiff abandons § 273 claim); *Sabasta v. Buckaroos, Inc.*, 2009 U.S. Dist. LEXIS 12754, *3-*2 (2009) (S.D. Iowa) (plaintiff briefly asserts § 273); *Powertech Tech., Inc. v. Tessera, Inc.*, 660 F.3d 1301, (Fed. Cir. 2011) (briefly discussing 273 in dicta).

[169] Carl Shapiro, *Prior User Rights*, 96 AMERICA ECON. REV. 92, 95 (2006).

[170] Stephen M. Maurer & Suzanne Scotchmer, *The Independent Invention Defense in Intellectual Property*, 69 ECONOMICA 535 (2002).

[171] Comments of Innovation Alliance, *supra* note 44.

[172] *Id.*

[173] Comments from Verizon Comm. Inc., and Google, Inc., to the USPTO, Comments of Verizon Comm. Inc., and Google, Inc. on the Study of Prior User Rights under the Leahy-Smith America Invents Act 3 (Nov. 8, 2011) [hereinafter "Comments from Google and Verizon"], http://www.uspto.gov/aia_implementation/pur-2011nov08- google_verizon.pdf.

[174] *Id.* at 6.

[175] Comments of Hospira Inc., *supra* note 43.

[176] NATIONAL INNOVATION SYSTEMS: A COMPARATIVE ANALYSIS (Richard Nelson ed. 1993).

[177] *See supra* notes 165-168 and accompanying text.

[178] *See* Kupferschmid, *supra* note 9, at 252.

[179] VAN EECKE, *supra* note 125, at 106.

[180] *See generally* Samuel Kortum & Josh Lerner, *Assessing the Contribution of Venture Capital to Innovation*, 31 RAND J. ECON. 674 (2000).

[181] Daniel Hoenig & Joachim Henkel, *Property Rights or Quality Signals? The Role of Patents in Venture Capital Financing*, 6th Annual Conf. of the EPIP Ass'n: Fine-Tuning IPR debates (2011), http://www.epip.eu/conferences/ epip06/papers/Parallel Session Papers/HOENIG Daniel.pdf.

[182] *See* Joel A.C. Baum & Brian S. Silverman, *Picking Winners or Building Them? Alliance, Intellectual, and Human Capital as Selection Criteria in Venture Financing and*

Performance of Biotechnology Startups, 19 J. BUS. VENTURING 411 (2003); Ronald J. Mann & Thomas W. Sager, *Patents, Venture Capital, and Software Start-Ups*, 36 RES. POL'Y 193 (2004).

[183] Comments from Tony Tether, to David Kappos et al., Undersecretary Dep. of Comm. for Intell. Prop., and Director of the USPTO 4 (Nov. 4, 2011) [hereinafter "Comments of Tony Tether"], http://www.uspto.gov/aia_implementation/pur-2011nov07-tony_tether.pdf.

[184] *See* Ellen Jalkut & Michael J. Remington, *Prior User Rights Expansion: Congress, the Courts, and Constitutional Considerations*, 82 BNA PAT., TRADEMARK, & COPYRIGHT J., Sept. 2, 2011, at 9, *available at* http://www.drinkerbiddle.com/files/ Publication/cbeafa8f-1b3b-4787-9fb8- 17739e59dca9/Presentation/PublicationAttachment/ 8ddaaed5-a6e9-4c30-b7af1b8ff12f0a7e/ITN_Remington_BNA.pdf.

[185] Comments from Brian Kahin, Senior Fellow, Computer & Comm. Indus. Ass'n, to the USPTO, Comments of the Computer & Commc'ns. Ind. Ass'n (Nov. 8, 2011) [hereinafter "Comments of the Computer & Commc'ns. Ind. Ass'n"] http://www.uspto.gov/aia_ implementation/pur-2011nov08-ccia.pdf.

[186] *See* VAN EECKE, *supra* note 125.

[187] *See* Cohen, *supra* note 166.

[188] Stuart J.H. Graham et al., *High Technology Entrepreneurs and the Patent System: Results of the 2008 Berkeley Patent Survey*, 24 BERKELEY TECH. L.J. 1255, 1305 (2009) (quoting a venture-capital investor).

[189] § 5, 125 Stat. at 292, 297-299 (2011).

[190] *See generally* David C. Mowery & Bhaven N. Sampat, *The Bayh-Dole Act of 1980 and University Industry Technology Transfer: A Model for Other OECD Governments?*, 30 J. TECH. TRANSFER 115 (2005).

[191] VAN EECKE, *supra* note 125.

[192] Comments of Microsoft Corp., *supra* note 43.

[193] Comments of Freischem, *supra* note 39.

[194] Comments from Various Higher Edu. Ass'n to the USPTO, Higher Educ. Comments on Prior User Rights 2 (Nov. 8, 2011) [hereinafter "Comments of Various Higher Educ. Ass'n"], http://www.uspto.gov/aia_implementation/pur-2011nov08-higher-education-associations. pdf.

[195] Comments from Carl E. Gulbrandsen, Managing Director, Wis. Alumni Research Found., to Elizabeth Shaw, IP Research Specialist Supervisor, USPTO, Office of Policy and External Affairs, In the Matter of Notice of Public Hearing and Request for Comments on Study of Prior User Rights 5-6 (Nov. 3, 2011) [hereinafter "Comments of Wis. Alumni Research Found."], http://www.uspto.gov/aia_implementation/pur-2011nov04- wisconsin_alumni_ research_foundation.pdf.

[196] Comments of Various Higher Educ. Ass'n, *supra* note 194.

[197] § 5, 125 Stat. at 292, 297-299.

[198] Comments of Various Higher Educ. Ass'n, *supra* note 194.

[199] *See* Comments of AIPLA, *supra* note 164.

[200] VAN EECKE, *supra* note 125, at 106.

[201] Kupferschmid, *supra* note 9, at 226.

[202] Comments from Paul Morgan, PTO Requested Comments on the AIA "Prior Commercial Use" Defense Legislation (Oct. 13, 2011) [hereinafter "Comments of Paul Morgan"], http://www.uspto.gov/aia_implementation/pur-2011oct13-paul_morgan.pdf.

[203] *Id.*

[204] *See generally*, Rochelle Cooper Dreyfuss, *Are Business Method Patents Bad for Business?*, 16 SANTA CLARA COMPUTER & HIGH TECH. L.J. 263 (2000).

[205] Edward L. MacCordy, *The Threat of Proposed Patent Law Changes to the Research University*, 20 J.C. & U.L. 295, 303 (1994).

[206] Chris P Konkol, *Prior-Invention Rights: The Excluded Middle*, 77 J. PAT. & TRADEMARK OFF. SOC'Y 666, 670 (1995); Kupferschmid, *supra* note 9, at 234.

[207] Kupferschmid, *supra* note 9, at 234.

[208] Comments of Various Higher Educ. Ass'n, *supra* note 194.

[209] *Id.*

[210] Comments of Verizon Communications Inc. and Google Inc., *supra* note 173.

[211] Comments of the Computer & Commc'ns. Ind. Ass'n, *supra* note 185.

[212] Comments of Greentech, *supra* note 47.

[213] Comments of Space Exploration Techs. Corp., *supra* note 47.

[214] Testimony of Thomas Kowalski, *supra* note 47; Gary Griswold, Consultant and Chair Emeritus of the Coalition for 21st Century Patent Reform, Comments at the Public Hearing on the Study of Prior User Rights 12-13 (Oct. 25, 2011) [hereinafter "Testimony of Gary Griswold"], http://www.uspto.gov/aia_implementation/20111025-prior-userrights-transcript.pdf.

[215] Comments of the Computer & Commc'ns. Ind. Ass'n, *supra* note 185.

[216] § 5, 125 Stat. at 297-299.

[217] Comments from Neil Thomas, to the USPTO, Prior User Rights 1 (Nov. 8, 2011) [hereinafter "Comments of Neil Thomas"], http://www.uspto.gov/aia_implementation/pur-2011nov08-neil_thomas.pdf.

[218] *See* Comments of Tony Tether, *supra* note 183.

[219] Jalkut, *supra* note 184.

[220] Comments of AIPLA, *supra* note 164.

[221] § 26, 125 Stat. at 338.

[222] U.S. CONST. art. I, § 8, cl. 8.

[223] Rebecca S. Eisenberg, *Patents and the Progress of Science: Exclusive Rights and Experimental Use*, 56 U. CHI. L. REV. 1017, 1022 (1989).

[224] *See, e.g.*, Comments of AIPLA, *supra* note 164, at 4 ("AIPLA does not believe that there are any legal or constitutional issues raised by the proposal to implement a prior inventor rights system, with the appropriate limitations as presently exist."); Testimony of Thomas Kowalski, *supra* note 47, at 15 ("The extent of the exclusivity of patent rights is among other things within Congress's discretion. . . . I do not see the prior user rights defense as granting any rights but only defining contours of the exclusivity granted by a patent."); Comments of the Computer & Commc'ns. Ind. Ass'n, *supra* note 185, at 3 ("Most arguments against prior user rights focus on the public disclosure function of the patent system—the principle that the patentee receives a limited monopoly in return for disclosing new knowledge to the public. Prior user rights do not change this basic *quid pro quo*."); *See generally* Comments at the Public Hearing on the Study of Prior User Rights (Oct. 25, 2011), http://www.uspto.gov/ aia_implementation/20111025-prior-user-rights-transcript.pdf; *and* Comments on the Study of Prior User Rights, *supra* note 33.

[225] Comments of AIPLA, *supra* note 164, at 4 (noting that the "U.S. Supreme Court addressed the compatibility of trade secret and patent laws in *Kewanee Oil Co. v. Bicron Corp.*, 416 U.S. 470 (1974), based upon the common goal of encouraging the development of new inventions.").

[226] Testimony of Thomas Kowalski, *supra* note 47 at 15-17 (noting that: (i) Congress has the power to define the right of exclusivity of a patent, and (ii) that a prior user rights defense existed in the U.S. prior to the 1952 Patent Act).

[227] *See, e.g.*, Comments from Google and Verizon, *supra* 173, at 3-6 ("[T]he courts and the PTO have long recognized that innovation is best served by giving companies the freedom 'to choose . . . the most commercially sound approach to commercially exploiting' their innovations."); Comments of Space Exploration Techs. Corp., *supra* note 47, at 2 ("At SpaceX, we believe that we must protect many of our innovations as trade secrets. . . . Due to well-known difficulties associated with enforcing patent rights in certain foreign jurisdictions, patents alone cannot adequately protect our technology. Moreover, patent specifications provide a roadmap to our unique technological advancements.").

[228] *See generally* Comments on the Study of Prior User Rights, *supra* note 33.

[229] *See, e.g.*, Comments of Wis. Alumni Research Found, *supra* note 195, at 2-5.

[230] *See id. and* Comments of Neil Thomas, *supra* note 217, at 2 ("The prior user rights provision seems to fly directly in the face of Art 1, Sec. 8, because the whole intent of the Founding Fathers was to encourage disclosure to society of new 'discoveries' by granting exclusive rights [...].").

[231] *See, e.g.*, Comments of Tony Tether, *supra* note 183, at 2 ("There is something fundamentally unfair about allowing something that is secret to erode a patent right.").

[232] U.S. CONST. art. I, § 8, cl. 8.

[233] *Graham v. John Deere Co.*, 383 U.S. 1, 5-6 (1966).

[234] *See Eldred v. Ashcroft*, 537 U.S. 186, 199-205 (2002) (considering the history of copyright terms in evaluating the constitutionality of copyright law).

[235] Act of 1839, ch. 88, § 7, 5 Stat. 353, 354 (repealed 1952) ("That every person or corporation who has, or shall have purchased or constructed any newly-invented machine, manufacture, or composition of matter, prior to the application by the inventor or discoverer of a patent, shall be held to possess the right to use and vend to others to be used, the specific machine, manufacture, or composition of matter, so made or purchased, without liability therefore to the inventor, or any other person interested in such invention; and no patent shall be held invalid by reason of such purchase, sale, or use prior to the application for a patent as aforesaid, except on proof of abandonment of such invention to the public, or that such purchase, sale, or prior use has been for more than two years prior to such application for a patent."); *see also McClurg v. Kingsland*, 42 U.S. 202, 208-09 (1843) ("The object of this provision is evidently twofold; first, to protect the person who has used the thing patented, by having purchased, constructed, or made the machine, etc., to which the invention is applied, from any liability to the patentee or his assignee. Second, to protect the rights, granted to the patentee, against any infringement by any other persons. . . . [I]t puts the person who has had such prior use on the same footing as if he had a special license from the inventor to use his invention; which, if given before the application for a patent, would justify the continued use after it issued without liability.").

[236] H.R. REP. NO. 1923, 82d Cong., Table 3, at 72 (1952).

[237] *See generally State Street Bank & Trust Co.*, 149 F.3d at 1368.

[238] American Inventor Protection Act of 1999, 35 U.S.C. § 273 (repealed and replaced by Leahy-Smith America Invents Act, § 5, 125 Stat. at 297-299).

[239] *See, e.g., Eldred*, 537 U.S. at 204-208.

[240] Cong. Rec. H4483 (daily ed. June 23, 2011) (statement of Rep. Lamar Smith).

[241] *Id.*

[242] *Id.* at 4492 (daily ed. June 23, 2011) (statements of Reps. Lamar Smith and Zoe Lofgren); Cong. Rec. S5426-27, 5430, 5540 (daily ed. Sept. 8, 2011) (statements of Sens. Roy Blunt, Patrick Leahy, Jon Kyle).

[243] Cong. Rec. H4483 (statement of Rep. Lamar Smith). *See also Eldred*, 537 U.S. at 205-206 (citing as a legitimate purpose the goal of Congress to harmonize United States copyright laws with European copyright laws "so that American authors would receive the same copyright protection in Europe as their European counterparts").

[244] *See* Kupferschmid, *supra* note 9, at 231 (addressing the argument that favoring trade secrets undermines the disclosure inducement theory underlying the patent system).

[245] *See* Comments of Wis. Alumni Research Found, *supra* note 195, at 2-3; Comments from P. Martin Simpson, Jr., Managing Counsel, Regents of the Univ. of Cal., to Elizabeth Shaw, IP Research Specialist Supervisor, USPTO, Office of Policy and External Affairs, Prior User Rights Written Comments 1 (Nov. 7, 2011), http://www.uspto.gov/aia_implementation/pur-2011nov07-university_of_ca.pdf; Comments of Tony Tether, *supra* note 183, at 3; Comments of Various Higher Educ. Ass'n, *supra* note 194, at 2.

[246] Comments of Wis. Alumni Research Found., *supra* note 195, at 3; *see also* Comments of Various Higher Educ. Ass'n, *supra* note 194, at 1 (arguing that prior user rights reduce certainty of patent rights, which impacts the ability of universities to license patents).

[247] *See* Comments from Google and Verizon, *supra* 173, at 9-10 (noting that due to circumstances in the high technology area, choosing trade secret protection over patent protection may make more economic sense); Comments from Coalition for Patent Fairness, *supra* note 47, at 2 (arguing that not every business has the resources to patent every development); Comments of Greentech, *supra* note 47, at 1-2 (explaining that because many innovations in this area tend to be manufacturing processes, reliance on trade secrets over patents is an important strategic tool).

[248] *See, e.g.*, Comments of the Computer & Commc'ns. Ind. Ass'n, *supra* note 185, at 3-4 (noting the substantial investment required to prepare and file patent applications, and the economic consequences of pursing an ideal of public disclosure over other forms of protection).

[249] *Id.* at 1-2; *see also* Comments of Space Exploration Techs. Corp., *supra* note 47, at 2 (noting difficulties in policing enforcement in foreign jurisdictions).

[250] *See* Comments from Coalition for Patent Fairness, *supra* note 47, at 2 (suggesting that the choice to file patent applications comes at a cost of decreased funding for basic research). Cost savings from choosing trade secret over patent protection can also contribute to a company's competitiveness in the field. *See* Comments of Space Exploration Techs. Corp., *supra* note 47, at 3-4.

[251] *See* Comments from Google and Verizon, *supra* 173, at 2, 3, 9 (noting the rapid pace and incremental nature of innovation in the high technology sector); Comments of the Computer & Commc'ns. Ind. Ass'n, *supra* note 185, at 3-4 (arguing that the faster and more continuous the pace of innovation, the more pressure would be placed on a company to file for patent protection, in the absence of prior user rights).

[252] *See, e.g.,* Comments from Google and Verizon, *supra* 173, at 1-2.

[253] Comments of the Computer & Commc'ns. Ind. Ass'n, *supra* note 185, at 4-5.

[254] 416 U.S. 470, 493 (1974).

[255] *See id.* at 491.

[256] *Id.*

[257] *See* Comments of Greentech, *supra* note 47, at 2, *and* Comments of Space Exploration Techs. Corp., *supra* note 47, at 2.

[258] Comments of Space Exploration Techs. Corp., *supra* note 47, at 3-4.

[259] Comments of Greentech, *supra* note 47, at 1-2.

[260] 35 U.S.C. § 102(g) (2006).

[261] The Philippines, the last country other than the United States to have a first-to-invent system, changed its law in 1998.

[262] *See* H.R. REP. NO. 112-98, at 40.

[263] 157 Cong. Rec. S5402, S5429-30 (Sept. 8, 2011) (Statement of Sen. Kyl) (noting that the expansion of prior user rights was "perhaps the most important" consequential change in view of the transition to first-inventor-to-file; *see also* 157 Cong. Rec. E1219, E1219 (June 28, 2011) (Extended Remarks of Rep. Smith) ("One key part of the transition [...] is the necessary inclusion of prior user rights under the new first-inventor-to-file system. The inclusion of prior user rights is essential to ensure that those who have invented and used a technology but choose not to disclose that technology [...] are provided a defense against someone who later patents the technology.").

[264] *Supra* note 263; *see also* 157 Cong. Rec. H4480, H4483 (June 23, 2011) (Statement of Rep. Smith) ("Prior-user rights are important as part of our change to a first-to-file system."); *and id.* at H4492 (Statement of Rep. Lofgren) ("[Y]ou cannot have first-to-file without robust prior-user rights.").

[265] Pub. L. No. 112-29, § 3, 125 Stat 284, 285-87 (2011) (replacing existing Section 102, including Section 102(g), with new provisions for determining novelty and non-obviousness based on filing dates instead of dates of "invention."); *see also* 157 Cong. Rec. S5402, S5426 (Sept. 8, 2011) (Statement of Sen. Blunt) (noting the elimination of the "prior invention bar to patentability under section 102(g)(2)").

[266] 35 U.S.C. § 102(g)(2).

[267] *See, e.g., Dunlop Holdings Ltd. v. Ram Golf Corp.*, 524 F.2d 33 (7th Cir. 1975). Although it is beyond the scope of this Report, it should be noted that there appears to be some ambiguity in the case law regarding the conditions under which a non-informing prior use by another may create patent-invalidating prior art under the existing statute. *Compare W.L. Gore & Assocs., Inc. v. Garlock, Inc.*, 721 F.2d 1540 (Fed. Cir. 1983) (non-informing use by a third party held not to invalidate the patent to another). The commentary on the subject is illustrative of the ambiguity. *Compare* Griswold & Ubel, *supra* note 9, at 572, n.20 (arguing that *Gore* was decided on grounds other than § 102(g)); *and* Jorda, *supra* note 15, at 28 (arguing that *Gore* only stands for the proposition that third party trade secrets are not invalidating prior art, which is entirely different from "a holding that the trade secret holder is an infringer vis-à-vis the patentee"); *with* Wendy H. Schacht & John R. Thomas, *Patent Reform in the 112th Congress: Innovation Issues*, Congressional Research Service Report for Congress, at 12 (June 30, 2011) (arguing that under the "settled patent law principle" announced in *Gore*, "[i]f an earlier inventor made secret commercial use of an invention, and another person independently invented the same technology later and obtained patent protection, then the trade secret holder could face liability for patent infringement").

[268] *See supra* note 20.

[269] *See, e.g.*, H.R. Rep. 112-98 at 87 (Letter from Secretary of Commerce Gary Locke to House Judiciary Committee Chairman Lamar Smith (May 31, 2011)) (providing a prior user rights defense is a "matter of fairness"); *see generally*, Kupferschmid, *supra* note 9 at 217, 229 (citing "fairness" as argument number one for and against prior user rights out a litany of arguments summarized in the article).

[270] Comments of Microsoft Corp., at 5.

[271] *See* Kyla Harriel, *Prior User Rights in a First-to-Invent Patent System: Why Not?*, 36 IDEA 543, 553-55 (1996) (noting the relationship between prior user rights and notions of

"economic justice" that underlie U.S. law as it relates to rewarding innovation and creativity).

[272] *See* Kupferschmid, *supra* note 9, at 217, 229.

[273] *See supra* note 20.

[274] *See id. and* Comments of Microsoft Corp., *supra* note 43, at 6.

[275] VAN EECKE *supra* note 125, at 23-24 (noting that prior user rights "are almost unanimously recognized in the [European Union] as being just and desirable on the grounds of both fairness and efficiency").

[276] *See, e.g.*, Comments from Freischem, *supra* note 39, at 3 (Germany); Comments of Telstra Corp., *supra* note 37, at 3(Australia); *and* Comments from Dr. Shoichi Okuyama, President, Japan Patent Attorneys Ass'n, to Mary Critharis et al., Senior Patent Counsel, USPTO, Office of Policy and External Affairs, Submission to the Request for Comments on the Study of Prior User Rights 1 (Nov. 7, 2011) [hereinafter "Comments from Japan Patent Attorneys Ass'n"] http://www.uspto.gov/aia_implementation/pur-2011nov08-jpaa.pdf.

[277] Comments from Freischem, *supra* note 39, at 3.

[278] *See* Lise Osterborg, *Towards a Harmonized Prior User Right Within a Common Market System*, 12 INT'L REV. INDUS. PROP. & COPYRIGHT 447, 456-59 (1981) (noting the dearth of reported prior user rights cases in France, England, Italy and the Netherlands); John Neukom, *A Prior Use Right for the Community Patent Convention,* 12 EUR. INTELL. PROP. REV. 165, 166 (1990) (few reported cases in West Germany); Comments of Microsoft Corp., *supra* note 43, at 4 (since 1977, there appear to be only 5 reported cases in the U.K.); *and* Comments of FICPI, *supra* note 2 (few reported cases in Europe and Brazil); *but see* Maeda, *supra* note 133, at 52 (reporting survey results suggesting frequent reliance on prior user rights by Japanese companies to avoid infringement litigation). A recent study on usage of the Section 102(g)(2) prior inventor defense in the U.S. suggests, by analogy, that prior user rights may be asserted with some frequency under the new regime provided in the AIA. Comments from Lex Machina, Inc., to the USPTO, U.S. Prior User Rights / Inventorship Study 1 (Nov. 7, 2011), http://www.uspto.gov/aia_implementation/pur-2011nov08-lex_machina.pdf (noting surprisingly high usage of Section 102(g)(2) in reported cases).

[279] Comments from Freischem, *supra* note 39, at 1-2; Comments of FICPI, *supra* note 40, at 2; Comments from Comments of Microsoft Corp., *supra* note 43, at 4 (referencing findings of the European Union report cited *supra*, note 275, including that settlement prior to litigation helped explain the low rate of reported usage of prior user rights in Europe); *see also* Comments from Coalition for Patent Fairness, at 4-5 (suggesting that lack of discovery in foreign jurisdictions limits application of the defense).

[280] Kupferschmid, *supra* note 9, at 221.

[281] Testimony of Gary Griswold, *supra* note 214, at 12-13; *see also*, Griswold & Ubel, *supra* note 9, at 577 (suggesting that lack of prior user rights in the United States may affect investment in U.S.-based assets); *and* Comments of Microsoft Corp., *supra* note 270, at 5 (noting that prior user rights permit the prior inventor/user to continue existing activities).

[282] Testimony of Gary Griswold, *supra* note 281.

[283] *See* Griswold & Ubel, *supra* note 9, at 568 (asserting that about 45% of U.S. patents granted in 1993 were foreign-owned). In terms of more recent available data, in Fiscal Year 2011, the USPTO issued 124,252 patents to residents of foreign countries out of a total of 244,430 patents issued in the same time period, suggesting that a little over 50% of patents issued in FY 2011 are foreign owned. *See* U.S. Patent and Trademark Office Performance and

Accountability Report Fiscal Year 2011, at 163, 165, and 168 (2011), *available at* http://www.uspto.gov/about/ stratplan/ar/2011/USPTOFY2011PAR.pdf.

[284] American Inventor Protection Act of 1999, 35 U.S.C. § 273 (1999) (repealed and replaced by the Leahy-Smith America Invents Act, Pub. L. No. 112-29, § 5, 125 Stat. 284, 297-99 (2011)).

[285] Leahy-Smith America Invents Act, Pub. L. No. 112-29, § 5, 125 Stat. 284, 297-99 (2011).

[286] *See* Notice of Public Hearing and Request for Comments on the Study of Prior User Rights, 76 Fed. Reg. 62,388 (Oct. 7, 2011).

[287] Greentech Industries consists of: Soraa, Inc.; Sakti3, Inc.; EchoFirst, Inc.; Solaria Corp.; Stion Corp.; Topanga Techs., Inc.

[288] Higher Education Associations consists of: Association of American Universities, American Council on Education, Association of American Medical Colleges, Association of Public and Land-grant Universities, Association of University Technology Managers, Council on Governmental Relations.

INDEX

D

E

F

G

H

I

J

K

L

M

S

T

U

V

W